BEYOND THE ATMOSP
AND IMPACT OF SP

MW01504836

AUTHOR

Jennifer Marlercurtis

&

Klaus Scherer

ISBN: 9798306325828

To all the pioneers of space weather research, whose relentless curiosity and dedication have paved the way for our understanding of the cosmos.

To our families and loved ones, whose unwavering support and encouragement have been the bedrock of our endeavors.

To the next generation of scientists and enthusiasts, may you continue to explore and expand the frontiers of our knowledge with the same passion and determination.

CHAPTER LIST

ACKNOWLEDGMENTS

Writing this book has been a journey of exploration and discovery, made possible by the support, guidance, and contributions of many individuals and organizations. We are deeply grateful to everyone who has played a part in bringing this work to fruition.

First and foremost, we would like to thank our families and loved ones. Your unwavering support and encouragement have been our bedrock throughout this endeavor. Your patience and understanding have been invaluable, allowing us to dedicate countless hours to research and writing.

We extend our deepest gratitude to our colleagues and peers in the field of space weather research. Your insights, discussions, and collaborations have significantly enriched the content of this book. We are particularly thankful to the members of the International Space Environment Service (ISES), the World Meteorological Organization (WMO), and the Committee on Space Research (COSPAR) for their pioneering work and for fostering a spirit of international cooperation.

A special thanks to the institutions and agencies that have supported our research and provided the resources necessary for this book. The National Aeronautics and Space Administration (NASA), the European Space Agency (ESA), and the National Oceanic and Atmospheric Administration (NOAA) have been instrumental in advancing our understanding of space weather through their comprehensive data and cutting-edge research.

We are grateful to our academic institutions for providing a stimulating environment and the resources needed to pursue this project. Your support has been crucial in facilitating our research and enabling us to contribute to the scientific community.

To our editors and publishers, your expertise and meticulous attention to detail have been invaluable in shaping this manuscript. Your dedication to excellence has greatly enhanced the quality of this book.

Space weather encompasses various phenomena originating from the Sun, which can influence the space environment and Earth's atmosphere. Understanding space weather is essential for mitigating its impacts on modern technology and infrastructure.

Historical Context of Space Weather

The study of space weather has a rich history, dating back to early observations of auroras and sunspots. The development of space weather science has been driven by technological advancements and the increasing reliance on space-based systems.

- **Early Observations:** Ancient civilizations recorded observations of auroras, attributing them to supernatural phenomena. The first scientific observations of sunspots were made by Galileo Galilei in the early 17th century.
- **Modern Era:** The advent of the space age in the mid-20th century revolutionized our understanding of space weather. The launch of satellites and space probes provided direct measurements of the Sun and its interactions with Earth's magnetosphere.

The Importance of Space Weather

Space weather affects various aspects of modern life, from satellite communications to power grid stability. Understanding these impacts is crucial for developing effective mitigation strategies.

- **Technological Dependence:** Modern society relies heavily on space-based technologies, including GPS, satellite communications, and weather forecasting. Space weather events can disrupt these systems, leading to significant economic and societal impacts.
- **Human Health:** Astronauts and airline passengers can be exposed to increased levels of radiation during space weather events. Protecting human health requires a thorough understanding of space weather phenomena and effective mitigation measures.

Components of Space Weather

Space weather is driven by dynamic processes on the Sun and its interactions with Earth's magnetosphere and atmosphere.

- **The Sun:** The primary source of space weather, the Sun emits a constant stream of charged particles known as the solar wind. Solar flares and coronal mass ejections (CMEs) are significant space weather events originating from the Sun.
- **Solar Wind:** The continuous flow of charged particles from the Sun interacts with Earth's magnetic field, causing various space weather phenomena.
- **Magnetosphere:** Earth's magnetic field acts as a shield, protecting the planet from the solar wind. However, this interaction can lead to geomagnetic storms and other space weather effects.

Space Weather Phenomena

Several key phenomena define space weather and its impacts on Earth and space-based systems.

- **Solar Flares:** Intense bursts of radiation from the Sun, solar flares can affect radio communications, navigation systems, and power grids.
- **Coronal Mass Ejections (CMEs):** Massive bursts of solar wind and magnetic fields rising above the solar corona, CMEs can cause geomagnetic storms when they collide with Earth's magnetosphere.
- **Geomagnetic Storms:** Disturbances in Earth's magnetosphere caused by solar wind and CMEs, geomagnetic storms can disrupt satellite operations, power grids, and communication systems.
- **Auroras:** Visible manifestations of space weather, auroras occur when charged particles from the solar wind interact with Earth's atmosphere.

Impacts on Technology and Society

The effects of space weather extend beyond the scientific community, impacting various sectors of modern society.

- **Satellite Operations:** Space weather can damage or disrupt satellites, affecting communication, navigation, and weather forecasting services.
- **Power Grids:** Geomagnetic storms can induce electrical currents in power lines, causing voltage instability and even blackouts.
- **Aviation:** Increased radiation levels during space weather events can pose risks to airline passengers and crew, particularly on polar routes.

Mitigation and Preparedness

Developing effective mitigation strategies is essential for protecting critical infrastructure and minimizing the impacts of space weather.

- **Forecasting and Monitoring:** Accurate space weather forecasting relies on continuous monitoring of the Sun and the space environment. Advanced models and observational data are crucial for predicting space weather events.
- **Engineering Solutions:** Designing resilient technologies and infrastructure can help mitigate the impacts of space weather. This includes hardening satellites against radiation and developing protective measures for power grids.

The Future of Space Weather Research

As our reliance on space-based systems grows, so does the importance of understanding and mitigating space weather impacts. Future research will focus on improving forecasting accuracy, developing more resilient technologies, and fostering international cooperation.

- **Emerging Technologies:** Advancements in sensor technology, data processing, and artificial intelligence will enhance our ability to monitor and predict space weather.
- **International Collaboration:** Global cooperation is essential for addressing the challenges of space weather. International organizations, such as the United Nations and the European Space Agency, play a critical role in coordinating space weather research and mitigation efforts.

Understanding space weather is vital for protecting modern technology and infrastructure from its potentially harmful effects. By studying the Sun and its interactions with Earth's magnetosphere, scientists can develop effective mitigation strategies and improve our resilience to space weather events. As we continue to explore and utilize space, the importance of space weather research will only grow, ensuring that we can navigate and thrive in this dynamic environment.

This introduction provides a comprehensive overview of space weather, its historical context, and its importance in modern society. The following chapters will delve deeper into each aspect of space weather, providing detailed information and analysis to enhance our understanding and preparedness.

The Sun, our nearest star, is a dynamic and complex celestial body that profoundly influences the space environment throughout the solar system. As the primary source of space weather, the Sun's activities, such as solar flares, coronal mass ejections (CMEs), and solar wind, have significant impacts on both space-based and terrestrial technologies. This chapter delves into the mechanisms of solar activity, its effects on space weather, and the implications for Earth's environment and technological systems.

2.1 Structure and Composition of the Sun

Core: The Sun's Energy Source The core is the innermost layer of the Sun, where nuclear fusion occurs. Hydrogen nuclei fuse to form helium, releasing vast amounts of energy in the process. This energy gradually moves outward through the Sun's layers.

- **Temperature:** Approximately 15 million degrees Celsius.
- **Density:** About 150 times the density of water.

Radiative Zone Surrounding the core is the radiative zone, where energy is transported outward by radiation. This region is characterized by dense plasma.

- **Temperature Range:** 2 million to 7 million degrees Celsius.
- **Radiation Transport:** Photons are absorbed and re-emitted by particles, a process that can take thousands of years for energy to pass through.

Convective Zone Above the radiative zone lies the convective zone, where energy is transported by convection currents. Hot plasma rises toward the surface, cools, and sinks back down.

- **Temperature Range:** 2 million degrees Celsius at the bottom to 5,500 degrees Celsius at the top.
- **Convection Currents:** These currents create granules visible on the Sun's surface.

Photosphere The photosphere is the visible surface of the Sun, from which sunlight is emitted.

- **Temperature:** Approximately 5,500 degrees Celsius.
- **Features:** Sunspots, which are cooler, dark areas caused by magnetic activity.

Chromosphere The chromosphere lies above the photosphere and is visible during solar eclipses as a reddish glow.

- **Temperature:** Ranges from 4,500 degrees Celsius near the photosphere to about 25,000 degrees Celsius at the top.
- **Features:** Prominences and spicules, which are jets of rising gas.

Corona The corona is the Sun's outer atmosphere, extending millions of kilometers into space.

- **Temperature:** Exceeds 1 million degrees Celsius.
- **Solar Wind Source:** The corona is the origin of the solar wind, a stream of charged particles flowing outward from the Sun.

2.2 Solar Activity and Space Weather

Solar Flares Solar flares are sudden, intense bursts of radiation caused by the release of magnetic energy stored in the Sun's atmosphere.

- **Classification:** Flares are classified based on their X-ray brightness in the wavelength range of 1 to 8 Angstroms: A, B, C, M, and X, with X being the most intense.
- **Impacts:** Flares can disrupt communications, navigation systems, and power grids on Earth. They can also pose radiation hazards to astronauts and high-altitude flights.

Coronal Mass Ejections (CMEs) CMEs are massive bursts of solar wind and magnetic fields rising above the solar corona.

- **Characteristics:** CMEs can release billions of tons of plasma and travel at speeds up to 3,000 kilometers per second.
- **Impacts:** When directed toward Earth, CMEs can cause geomagnetic storms, affecting satellite operations, power grids, and GPS accuracy.

Solar Wind The solar wind is a continuous flow of charged particles from the Sun, primarily electrons and protons.

- **Components:** The solar wind consists of the slow solar wind (approximately 400 km/s) and the fast solar wind (up to 800 km/s).
- **Interaction with Earth:** The solar wind interacts with Earth's magnetosphere, causing phenomena such as auroras and geomagnetic storms.

Sunspots and Solar Cycles Sunspots are temporary phenomena on the Sun's photosphere that appear as dark spots compared to surrounding areas.

- **Magnetic Activity:** Sunspots are regions of intense magnetic activity and are cooler than the surrounding photosphere.
- **Solar Cycles:** The number of sunspots varies in an approximately 11-year cycle, known as the solar cycle. The cycle includes periods of solar maximum (high sunspot activity) and solar minimum (low sunspot activity).

2.3 Mechanisms of Solar Weather Events

Magnetic Reconnection Magnetic reconnection is a process where magnetic field lines from different magnetic domains are forced together and rearrange, releasing a significant amount of energy.

- **Occurrence:** Typically occurs in the Sun's corona, leading to solar flares and CMEs.
- **Impact:** Drives the most energetic solar events, influencing space weather conditions.

Solar Wind Interaction with Earth's Magnetosphere The interaction of the solar wind with Earth's magnetosphere is a dynamic process that shapes space weather effects.

- **Bow Shock:** The point where the solar wind slows down and deflects around Earth, forming a bow shock.
- **Magnetopause:** The boundary between the solar wind and Earth's magnetosphere.
- **Magnetotail:** The elongated region of the magnetosphere on the side opposite the Sun, created by the pressure of the solar wind.

Auroras Auroras, also known as the Northern and Southern Lights, are visual manifestations of space weather.

- **Formation:** Caused by the interaction of solar wind particles with Earth's magnetic field and atmosphere.
- **Colors:** Different gases in Earth's atmosphere emit various colors when excited by solar particles (e.g., green from oxygen, red from nitrogen).

2.4 Solar Observation and Monitoring

Ground-Based Observatories Ground-based solar observatories provide continuous monitoring of the Sun's activity.

- **Telescopes:** Equipped with filters to observe different wavelengths of light, including visible, ultraviolet, and X-rays.
- **Helioseismology:** The study of the Sun's interior through observations of surface oscillations.

Space-Based Instruments Satellites and space probes offer an unobstructed view of the Sun and space weather phenomena.

- **SOHO (Solar and Heliospheric Observatory):** Provides detailed observations of the Sun's atmosphere and solar wind.
- **SDO (Solar Dynamics Observatory):** Captures high-resolution images of the Sun in multiple wavelengths.
- **ACE (Advanced Composition Explorer):** Measures the composition of the solar wind and cosmic rays.

Data Analysis and Modeling Advanced data analysis and modeling techniques are essential for understanding and predicting space weather events.

- **Simulation Models:** Help simulate solar wind, magnetic reconnection, and other space weather phenomena.
- **Forecasting Tools:** Provide predictions of solar flares, CMEs, and their potential impacts on Earth.

2.5 Impacts of Solar Activity on Earth

Geomagnetic Storms Geomagnetic storms are disturbances in Earth's magnetosphere caused by solar wind and CMEs.

- **Effects on Power Grids:** Can induce geomagnetically induced currents (GICs) that affect transformers and other components of power grids.
- **Satellite Operations:** Can disrupt satellite communication, navigation, and operations due to increased radiation and charged particles.

Radiation Hazards Solar radiation poses risks to astronauts, aviation, and electronic systems.

- **Astronaut Exposure:** Increased radiation levels during solar events can be harmful to astronauts on the International Space Station (ISS) and future deep-space missions.
- **Aviation:** High-altitude flights, particularly near the poles, can experience increased radiation exposure during solar storms.

Technological Disruptions Solar activity can disrupt various technological systems on Earth.

- **Communication Systems:** HF radio communications can be affected by solar flares and ionospheric disturbances.
- **GPS Navigation:** Accuracy of GPS signals can be degraded by space weather events, affecting navigation and timing applications.

Climate and Weather Solar variability can influence Earth's climate and weather patterns.

- **Solar Irradiance:** Changes in the Sun's energy output can affect global temperatures and climate patterns.
- **Weather Patterns:** Some studies suggest a link between solar activity and weather phenomena, though these connections are still being researched.

2.6 Mitigation Strategies for Space Weather Impacts

Engineering Solutions Designing resilient systems can mitigate the impacts of space weather.

- **Hardened Satellites:** Using radiation-hardened components to protect satellites from solar radiation.
- **Power Grid Protection:** Implementing protective measures such as GIC blocking devices to safeguard power grids.

Operational Strategies Operational strategies can minimize the risks of space weather events.

- **Satellite Operations:** Adjusting satellite orbits and operations during space weather events to minimize damage.
- **Aviation Procedures:** Rerouting high-altitude flights away from polar regions during solar storms to reduce radiation exposure.

International Cooperation Global collaboration is essential for effective space weather mitigation.

- **Data Sharing:** Sharing space weather data and forecasts among countries and organizations.
- **Joint Research:** Collaborating on space weather research and developing common mitigation strategies.

2.7 Future Research and Developments

Advancements in Observation Technology Improving observation technology will enhance our understanding of the Sun and space weather.

- **Next-Generation Telescopes:** Developing advanced telescopes with higher resolution and sensitivity.
- **Interplanetary Probes:** Launching probes to study the Sun and space weather phenomena from different vantage points.

Improved Modeling and Prediction Advances in modeling and prediction will improve space weather forecasts.

- **Machine Learning:** Using artificial intelligence and machine learning to analyze space weather data and improve predictions.
- **Integrated Models:** Developing comprehensive models that integrate observations, simulations, and theoretical frameworks.

Human Exploration and Space Weather As human space exploration expands, understanding and mitigating space weather will be crucial.

- **Lunar and Mars Missions:** Planning for radiation protection and the impacts of solar activity on future lunar bases and Mars habitats is essential. This involves developing advanced shielding materials and real-time monitoring systems to protect astronauts from harmful solar radiation.

- **Deep Space Travel:** As missions venture beyond the Earth-Moon system, the need for robust space weather forecasting and mitigation strategies becomes even more critical. This includes understanding how the interstellar medium and solar wind affect spacecraft traveling to the outer planets and beyond.

Advancements in Observation Technology Improving observation technology will enhance our understanding of the Sun and space weather.

- **Next-Generation Telescopes:** Developing advanced telescopes with higher resolution and sensitivity, such as the Parker Solar Probe and the upcoming Solar Orbiter mission, which will provide unprecedented close-up views of the Sun's surface and its magnetic activity.
- **Interplanetary Probes:** Launching probes to study the Sun and space weather phenomena from different vantage points, such as the Solar and Heliospheric Observatory (SOHO) and the upcoming missions by NASA and ESA.

Improved Modeling and Prediction Advances in modeling and prediction will improve space weather forecasts.

- **Machine Learning:** Using artificial intelligence and machine learning to analyze space weather data and improve predictions. These technologies can help identify patterns in solar activity and predict space weather events with greater accuracy.
- **Integrated Models:** Developing comprehensive models that integrate observations, simulations, and theoretical frameworks. These models can simulate the interactions between the solar wind, magnetosphere, and ionosphere to provide more accurate space weather forecasts.

International Collaboration and Policy Development International cooperation is crucial for addressing the global impacts of space weather.

- **Policy Development:** Collaborating on the development of international policies and guidelines for space weather monitoring and mitigation. Organizations such as the International Space Weather Initiative (ISWI) and the Space Weather Prediction Center (SWPC) play vital roles in coordinating these efforts.
- **Global Data Sharing:** Promoting the sharing of space weather data among nations and institutions to enhance global preparedness. This includes real-time data from solar observatories and space-based monitoring systems.

Education and Public Awareness Raising awareness about space weather and its impacts is essential for fostering a resilient society.

- **Educational Programs:** Implementing educational programs at all levels to teach students about space weather and its effects on technology and daily life. Universities and research institutions can offer specialized courses in space weather and space physics.
- **Public Outreach:** Conducting public outreach campaigns to inform the general public about space weather phenomena and preparedness measures. This can include workshops, seminars, and media campaigns to raise awareness about the importance of space weather research.

New Technologies and Innovations The development of new technologies will enhance our ability to monitor and mitigate space weather impacts.

- **Advanced Shielding Materials:** Researching and developing new materials for spacecraft and habitats that provide better protection against solar radiation and cosmic rays.
- **Space Weather Instruments:** Designing and deploying advanced instruments for measuring and analyzing space weather phenomena, such as magnetometers, particle detectors, and solar imagers.

Understanding the Sun and its role in space weather is crucial for protecting our technological infrastructure and ensuring the safety of human space exploration. This chapter has provided a comprehensive overview of the Sun's structure, the mechanisms driving solar activity, and the impacts of space weather on Earth and space-based systems. As we continue to explore the solar system and rely more on space-based technologies, the need for advanced space weather forecasting and mitigation strategies will only grow. By investing in research, international collaboration, and public education, we can better prepare for and mitigate the effects of space weather, ensuring a safer and more resilient future.

Key Takeaways

1. **The Sun's Structure and Activity:** Understanding the Sun's layers and processes is essential for comprehending space weather phenomena.
2. **Solar Flares and CMEs:** These significant solar events can have profound impacts on Earth's technology and environment.
3. **Magnetic Reconnection:** A critical process driving solar activity and influencing space weather.

4. **Impacts on Technology and Health:** Space weather affects satellite operations, power grids, communication systems, and human health.
5. **Mitigation Strategies:** Engineering solutions, operational strategies, and international cooperation are key to mitigating space weather impacts.
6. **Future Research:** Advancements in observation technology, modeling, and prediction will enhance our ability to forecast and respond to space weather events.
7. **Public Awareness:** Education and outreach are crucial for raising awareness about space weather and its effects.

This comprehensive chapter on the Sun and its role in space weather sets the stage for understanding the subsequent chapters, which will explore specific space weather phenomena, their impacts on technology, and strategies for mitigation and preparedness.

CHAPTER 3: SOLAR WIND AND ITS EFFECTS ON THE EARTH

3.1 Introduction to Solar Wind

Definition and Nature of Solar Wind

The solar wind is a continuous flow of charged particles—mainly electrons, protons, and alpha particles—emitted by the Sun's corona. These particles carry with them a portion of the Sun's magnetic field, known as the interplanetary magnetic field (IMF). The solar wind plays a crucial role in shaping the space environment throughout the solar system and has profound effects on Earth's magnetosphere, ionosphere, and technological systems.

Discovery and Historical Context

The concept of solar wind was first proposed by the German physicist Ludwig Biermann in the 1950s, based on observations of comet tails that always pointed away from the Sun. The existence of solar wind was later confirmed by spacecraft observations, notably by the Soviet Luna 1 mission in 1959 and subsequent measurements by NASA's Mariner 2 spacecraft in 1962.

Properties of Solar Wind

- **Speed and Density:** The solar wind has an average speed of about 400 km/s, but it can vary between 300 km/s and 800 km/s. The density of the solar wind typically ranges from 1 to 10 particles per cubic centimeter.
- **Temperature:** The temperature of the solar wind varies, but it generally ranges from 1 million to 2 million Kelvin.
- **Magnetic Field:** The interplanetary magnetic field (IMF) carried by the solar wind has an average strength of about 5 nanoTesla near Earth.

3.2 Origin and Acceleration of Solar Wind

Source Regions in the Sun

The solar wind originates from the outer layer of the Sun, the corona. There are two primary types of solar wind: the fast solar wind and the slow solar wind.

- **Fast Solar Wind:** Emanates from coronal holes, which are regions of the corona with open magnetic field lines. These regions allow particles to escape easily into space, resulting in a high-speed wind (approximately 800 km/s).

- **Slow Solar Wind:** Originates from the streamer belt around the Sun's equator, where the magnetic field lines are closed and the wind escapes through a more gradual process, resulting in a slower speed (approximately 400 km/s).

Mechanisms of Acceleration

The exact mechanisms that accelerate the solar wind are still subjects of ongoing research, but several theories have been proposed:

- **Wave-Particle Interactions:** Alfvén waves and other magnetohydrodynamic (MHD) waves generated in the corona can transfer energy to particles, accelerating them to high speeds.
- **Magnetic Reconnection:** The process of magnetic reconnection in the corona can release significant amounts of energy, accelerating particles and contributing to the solar wind.

Observations and Models

- **Spacecraft Observations:** Missions like the Solar and Heliospheric Observatory (SOHO), Parker Solar Probe, and Solar Orbiter have provided detailed observations of the solar wind and its acceleration regions.
- **Theoretical Models:** Computational models, such as magnetohydrodynamic (MHD) simulations, help researchers understand the complex processes involved in the acceleration and propagation of the solar wind.

3.3 Interaction of Solar Wind with Earth's Magnetosphere

Structure of Earth's Magnetosphere

Earth's magnetosphere is a dynamic region where the planet's magnetic field interacts with the solar wind. It acts as a protective shield, deflecting most of the solar wind particles around the planet.

- **Bow Shock:** The point where the solar wind slows down abruptly upon encountering Earth's magnetosphere, forming a shock wave.
- **Magnetopause:** The boundary between the solar wind and Earth's magnetosphere, where the pressure from the solar wind is balanced by Earth's magnetic field.

- **Magnetotail:** The elongated region of the magnetosphere on the side opposite the Sun, formed by the pressure of the solar wind stretching Earth's magnetic field lines.

Dynamic Processes

- **Magnetic Reconnection:** Occurs on the dayside magnetopause and in the magnetotail, allowing solar wind energy to enter the magnetosphere. This process is crucial for driving geomagnetic storms and substorms.
- **Plasma Transfer:** Solar wind plasma can enter the magnetosphere through processes such as magnetic reconnection, Kelvin-Helmholtz instabilities, and direct entry through the polar cusps.

Effects on Earth's Magnetosphere

- **Geomagnetic Storms:** Large-scale disturbances in Earth's magnetosphere caused by enhanced solar wind and CMEs. These storms can last for several days and have significant impacts on technological systems.
- **Auroras:** Visible manifestations of solar wind particles interacting with Earth's atmosphere, creating light displays in the polar regions known as the Northern and Southern Lights.

3.4 Ionospheric and Atmospheric Effects

Ionospheric Responses

The ionosphere, a layer of Earth's atmosphere ionized by solar radiation, is highly responsive to changes in solar wind conditions.

- **Disturbances:** Increased solar wind activity can lead to ionospheric disturbances, affecting radio wave propagation and GPS signals.
- **Total Electron Content (TEC):** The TEC of the ionosphere can increase dramatically during geomagnetic storms, affecting satellite communications and navigation.

Atmospheric Heating

The interaction of the solar wind with Earth's magnetosphere can also lead to heating of the upper atmosphere.

- **Joule Heating:** Caused by enhanced electric currents in the ionosphere during geomagnetic storms.
- **Neutral Density Increase:** Heating of the upper atmosphere can increase the density of neutral particles, affecting satellite drag and orbital decay.

3.5 Impacts on Technology and Human Activity

Satellite Operations

Solar wind and geomagnetic storms can have various effects on satellite operations:

- **Radiation Damage:** Increased radiation can damage satellite electronics and degrade solar panels.
- **Surface Charging:** Differential charging of satellite surfaces can lead to electrostatic discharges, potentially damaging components.
- **Communication Disruptions:** Increased ionospheric activity can disrupt satellite communications and GPS signals.

Power Grids

Geomagnetic storms can induce currents in power lines, leading to potential disruptions in power grids:

- **Geomagnetically Induced Currents (GICs):** Can cause transformer overheating, voltage instability, and, in severe cases, power outages.
- **Historical Events:** The March 1989 geomagnetic storm caused a major blackout in Quebec, Canada, affecting millions of people.

Aviation and Human Health

- **Aviation:** Increased radiation levels during solar events can pose risks to high-altitude flights, particularly on polar routes. Airlines may need to reroute flights to avoid exposure.
- **Human Health:** Astronauts in space and passengers on high-altitude flights are at risk of increased radiation exposure during solar events. Protective measures are essential to mitigate these risks.

3.6 Monitoring and Forecasting Solar Wind

Space-Based Observatories

Numerous spacecraft monitor the Sun and solar wind to provide real-time data and improve forecasting capabilities:

- **SOHO (Solar and Heliospheric Observatory):** Provides continuous observations of the Sun's corona and solar wind.
- **ACE (Advanced Composition Explorer):** Monitors the solar wind and cosmic rays near the Lagrange point L1.
- **Parker Solar Probe:** Studies the solar corona and the processes that accelerate the solar wind.

Ground-Based Observatories

Ground-based observatories complement space-based observations by monitoring solar activity and its effects on Earth:

- **Magnetometers:** Measure variations in Earth's magnetic field, providing data on geomagnetic activity.
- **Ionosondes:** Measure the ionosphere's density and composition, helping to assess the impact of solar wind on communications.

Forecasting Models

Advancements in modeling and simulation have improved the ability to forecast solar wind and space weather events:

- **MHD Simulations:** Magnetohydrodynamic models simulate the behavior of the solar wind and its interactions with Earth's magnetosphere.
- **Empirical Models:** Based on historical data, these models predict the likelihood and severity of geomagnetic storms and other space weather phenomena.

3.7 Mitigation Strategies

Technological Solutions

Developing resilient technologies is crucial for mitigating the impacts of solar wind and space weather:

- **Satellite Design:** Using radiation-hardened components and implementing protective shielding to reduce damage from solar radiation.

- **Power Grid Protection:** Installing GIC blocking devices and enhancing grid infrastructure to withstand geomagnetic storms.

Operational Strategies

Adjusting operational procedures can help minimize the risks posed by space weather events:

- **Satellite Operations:** Adjusting satellite orbits and operations during geomagnetic storms to minimize exposure to radiation.
- **Aviation Procedures:** Rerouting high-altitude flights during solar events to reduce radiation exposure for passengers and crew.

International Collaboration

Global cooperation is essential for effective space weather mitigation:

- **Data Sharing:** Promoting the exchange of space weather data and forecasts among countries and organizations.
- **Joint Research Initiatives:** Collaborating on research projects to improve understanding and mitigation of space weather impacts.

3.8 Future Research Directions

Advancements in Solar Wind Research

Ongoing research aims to enhance our understanding of the solar wind and its interactions with Earth's environment:

- **New Missions:** Upcoming missions, such as ESA's Solar Orbiter and NASA's IMAP (Interstellar Mapping and Acceleration Probe), will provide new insights into the solar wind and its effects.
- **High-Resolution Observations:** Developing instruments with higher resolution and sensitivity to capture detailed observations of solar wind phenomena.

Improved Forecasting Techniques

Advances in forecasting techniques will improve our ability to predict space weather events and mitigate their impacts:

- **Machine Learning and AI:** Using artificial intelligence to analyze space weather data and improve prediction accuracy.
- **Integrated Models:** Developing comprehensive models that integrate solar, magnetospheric, and ionospheric data for more accurate forecasts.

Understanding Long-Term Trends

Studying long-term trends in solar activity and space weather is essential for predicting future conditions and planning mitigation strategies:

- **Solar Cycle Variability:** Investigating the causes of variability in the solar cycle and its impacts on space weather.
- **Historical Data Analysis:** Analyzing historical data from past solar cycles to identify patterns and improve forecasting models.

3.9 Case Studies of Solar Wind Impacts

The Carrington Event (1859)

One of the most significant geomagnetic storms on record, the Carrington Event, caused widespread disruptions in telegraph systems and produced vivid auroras visible at low latitudes.

- **Cause:** A powerful CME struck Earth, inducing strong geomagnetic storms.
- **Impact:** Telegraph systems experienced surges and outages, and auroras were seen as far south as the Caribbean.

The March 1989 Geomagnetic Storm

This event caused a major blackout in Quebec, Canada, affecting millions of people and highlighting the vulnerability of power grids to space weather.

- **Cause:** A CME resulted in intense geomagnetic storms and geomagnetically induced currents (GICs).
- **Impact:** The Hydro-Québec power grid collapsed, leaving 6 million people without power for nine hours.

The Halloween Storms (2003)

A series of intense solar storms in October-November 2003 disrupted satellite operations, communication systems, and power grids worldwide.

- **Cause:** Multiple CMEs and solar flares led to severe geomagnetic activity.
- **Impact:** Satellite anomalies, aviation disruptions, and power grid issues were reported globally.

3.10 Technological Advancements and Space Weather Resilience

Next-Generation Satellites

Developing next-generation satellites with enhanced resilience to space weather is critical for ensuring the continuity of space-based services.

- **Radiation-Hardened Electronics:** Incorporating radiation-hardened components to withstand high-energy particle radiation.
- **Real-Time Monitoring:** Equipping satellites with instruments to monitor space weather conditions in real-time and adjust operations accordingly.

Smart Grid Technology

Implementing smart grid technology can improve the resilience of power grids to geomagnetic storms.

- **GIC Blocking Devices:** Installing devices to block or divert geomagnetically induced currents.
- **Adaptive Grid Management:** Using real-time data and AI to dynamically manage power distribution and mitigate the effects of geomagnetic disturbances.

Aviation Safety Measures

Enhancing aviation safety measures to protect passengers and crew from increased radiation exposure during solar events is crucial.

- **Flight Path Adjustments:** Implementing procedures to reroute flights away from polar regions during solar storms.
- **Radiation Monitoring:** Equipping aircraft with radiation monitoring instruments to assess exposure levels and take appropriate action.

3.11 International Policy and Cooperation

Global Space Weather Programs

International cooperation is essential for effective space weather monitoring and mitigation.

- **International Space Weather Initiative (ISWI):** A global program that promotes space weather research, education, and data sharing among participating countries.
- **World Meteorological Organization (WMO):** The WMO collaborates with space weather agencies to integrate space weather forecasting into global meteorological services.

Standardizing Space Weather Data

Developing international standards for space weather data collection, analysis, and dissemination can enhance global preparedness.

- **ISO Standards:** Implementing ISO standards for space weather data to ensure consistency and reliability across different regions and organizations.
- **Global Data Sharing Platforms:** Establishing platforms for sharing real-time space weather data and forecasts among countries and institutions.

3.12 Educational and Public Awareness Initiatives

Space Weather Education Programs

Educational programs can raise awareness about space weather and its impacts on technology and daily life.

- **School Curricula:** Integrating space weather topics into school curricula to educate students about solar activity and its effects.
- **University Courses:** Offering specialized courses in space weather and space physics at universities to train the next generation of scientists and engineers.

Public Outreach and Engagement

Public outreach initiatives can inform the general public about space weather and encourage preparedness.

- **Workshops and Seminars:** Hosting workshops and seminars to educate communities and industry stakeholders about space weather impacts and mitigation strategies.
- **Media Campaigns:** Using media campaigns to raise awareness about space weather and promote protective measures.

The solar wind is a fundamental aspect of space weather, with far-reaching effects on Earth's magnetosphere, ionosphere, and technological systems. Understanding the mechanisms of solar wind generation, its interaction with Earth's magnetic field, and the resulting impacts on technology and human activities is crucial for developing effective mitigation strategies.

This chapter has provided a comprehensive overview of the solar wind, including its origin, properties, and effects on Earth. By exploring case studies of significant space weather events and discussing advancements in technology and forecasting, we have highlighted the importance of preparedness and resilience in the face of space weather challenges.

As we look to the future, continued research, international cooperation, and public awareness will be essential for addressing the complexities of space weather and ensuring the safety and reliability of our technological infrastructure. By investing in education, innovation, and global collaboration, we can better prepare for and mitigate the impacts of solar wind and other space weather phenomena.

CHAPTER 4: MAGNETOSPHERE: EARTH'S SHIELD

The Earth's magnetosphere is a complex and dynamic region of space where the planet's magnetic field interacts with the solar wind. This interaction creates a protective shield that deflects most of the solar wind particles and cosmic rays, thereby safeguarding life on Earth and maintaining the stability of technological systems in space and on the ground. This chapter delves into the detailed structure and dynamics of the magnetosphere, its interaction with solar wind, and its critical role in protecting Earth.

4.1 Introduction to the Magnetosphere

Definition and Importance

The magnetosphere is a region around the Earth where the geomagnetic field dominates over the solar wind's dynamic pressure. It extends from the ionosphere to tens of Earth radii into space and plays a crucial role in shielding the Earth from solar and cosmic radiation.

Discovery and Historical Context

The concept of the magnetosphere emerged in the mid-20th century, with the term "magnetosphere" being coined by Thomas Gold in 1959. The discovery of the Van Allen radiation belts by the Explorer 1 and Explorer 3 missions in 1958 was a significant milestone in understanding the magnetosphere's structure and function.

4.2 Structure of the Magnetosphere

Magnetosphere Boundaries

- **Bow Shock:** The boundary where the solar wind slows down abruptly upon encountering Earth's magnetosphere, forming a shock wave. The bow shock stands about 90,000 km from Earth on the sunward side.
- **Magnetopause:** The outer boundary of the magnetosphere where the pressure from the solar wind is balanced by Earth's magnetic field. It typically lies at about 10 Earth radii (RE) from the Earth on the dayside but can vary based on solar wind conditions.

- **Magnetotail:** The elongated region of the magnetosphere on the night side of Earth, stretched by the solar wind, extending up to 1000 RE away from Earth.

Regions within the Magnetosphere

- **Plasmasphere:** A torus-shaped region of cold, dense plasma that co-rotates with the Earth, extending up to 4-5 RE.
- **Radiation Belts:** The Van Allen radiation belts consist of energetic charged particles trapped by Earth's magnetic field. The inner belt extends from about 1 to 3 RE, and the outer belt from about 3 to 10 RE.
- **Plasma Sheet:** Located in the magnetotail, it is a region of hot, dense plasma that plays a significant role in magnetospheric dynamics.

Magnetic Field Lines

- **Closed Field Lines:** Near the Earth, magnetic field lines are closed, looping back to the surface.
- **Open Field Lines:** In the polar regions, some magnetic field lines are open, extending into space and connecting to the interplanetary magnetic field (IMF).

4.3 Dynamics of the Magnetosphere

Magnetic Reconnection

Magnetic reconnection is a process where magnetic field lines from different magnetic domains are forced together, break, and reconnect, releasing significant amounts of energy.

- **Occurrence:** Reconnection can occur on the dayside magnetopause and in the magnetotail.
- **Impact:** Drives geomagnetic storms and substorms, injecting energy into the magnetosphere and ionosphere.

Solar Wind Interaction

The interaction between the solar wind and the magnetosphere drives many of the dynamic processes within the magnetosphere.

- **Compression and Expansion:** The magnetosphere compresses during periods of high solar wind pressure and expands during low pressure.

- **Energy Transfer:** Solar wind energy is transferred to the magnetosphere through processes such as magnetic reconnection and Kelvin-Helmholtz instabilities.

Auroral Phenomena

Auroras are visible manifestations of the interaction between the solar wind and Earth's magnetosphere, occurring mainly in the polar regions.

- **Formation:** Caused by the precipitation of energetic particles from the magnetosphere into the upper atmosphere.
- **Colors and Shapes:** Different gases in the atmosphere emit various colors when excited by these particles. Auroras can appear as curtains, arcs, or spirals.

4.4 Magnetosphere-Ionosphere Coupling

Field-Aligned Currents

Field-aligned currents (also known as Birkeland currents) flow along magnetic field lines between the magnetosphere and the ionosphere.

- **Formation:** Driven by pressure gradients and electric fields in the magnetosphere.
- **Impact:** They transfer energy and momentum between the magnetosphere and ionosphere, playing a crucial role in auroral dynamics.

Ionospheric Conductivity

The ionosphere's conductivity influences the coupling between the magnetosphere and ionosphere.

- **Day-Night Variation:** Conductivity is higher during the day due to solar radiation ionizing the atmosphere.
- **Seasonal and Solar Cycle Variation:** Varies with the seasons and the 11-year solar cycle.

Joule Heating

Joule heating occurs when electric currents flow through the ionosphere's conductive layer, converting kinetic energy into thermal energy.

- **Impact:** Causes localized heating of the upper atmosphere, leading to changes in atmospheric density and dynamics.

4.5 Impacts of Magnetospheric Disturbances

Geomagnetic Storms

Geomagnetic storms are large-scale disturbances in Earth's magnetosphere caused by enhanced solar wind and CMEs.

- **Classification:** Geomagnetic storms are classified based on the disturbance storm time (Dst) index, with storms categorized as weak, moderate, strong, or severe.
- **Effects on Technology:** Can induce geomagnetically induced currents (GICs) in power grids, disrupt satellite operations, and affect communication systems.

Substorms

Substorms are shorter, more localized disturbances in the magnetosphere, often occurring within geomagnetic storms.

- **Phases:** Substorms have three phases: growth, expansion, and recovery.
- **Impact:** Substorms contribute to the injection of energetic particles into the radiation belts and the formation of auroras.

Radiation Belt Dynamics

The Van Allen radiation belts are highly dynamic, with particle populations varying in response to geomagnetic activity.

- **Acceleration Mechanisms:** Wave-particle interactions, such as chorus and hiss waves, can accelerate particles to high energies.
- **Loss Mechanisms:** Particles can be lost from the belts through interactions with the atmosphere or by being swept away by solar wind.

4.6 Monitoring and Forecasting Magnetospheric Activity

Ground-Based Observatories

Ground-based observatories play a critical role in monitoring magnetospheric activity.

- **Magnetometers:** Measure variations in Earth's magnetic field, providing data on geomagnetic activity.
- **Ionosondes:** Measure ionospheric properties, helping to assess the impact of magnetospheric disturbances on communications.

Space-Based Missions

Space-based missions provide crucial data on the structure and dynamics of the magnetosphere.

- **Cluster Mission:** A constellation of four ESA spacecraft studying the small-scale structure and dynamics of the magnetosphere.
- **THEMIS Mission:** NASA's mission to investigate the triggers of substorms and their effects on the magnetosphere.
- **Van Allen Probes:** Studied the radiation belts and provided insights into particle acceleration and loss processes.

Data Analysis and Modeling

Advanced data analysis and modeling techniques are essential for understanding and predicting magnetospheric activity.

- **Simulation Models:** Global MHD simulations and kinetic models help researchers study the complex processes within the magnetosphere.
- **Empirical Models:** Based on historical data, these models predict the likelihood and severity of geomagnetic storms and substorms.

4.7 Mitigation Strategies for Magnetospheric Impacts

Technological Solutions

Developing resilient technologies is crucial for mitigating the impacts of magnetospheric disturbances.

- **Satellite Design:** Using radiation-hardened components and implementing protective shielding to reduce damage from energetic particles.
- **Power Grid Protection:** Installing GIC blocking devices and enhancing grid infrastructure to withstand geomagnetic storms.

Operational Strategies

Adjusting operational procedures can help minimize the risks posed by magnetospheric disturbances.

- **Satellite Operations:** Adjusting satellite orbits and operations during geomagnetic storms to minimize exposure to radiation.
- **Aviation Procedures:** Rerouting high-altitude flights during solar events to reduce radiation exposure for passengers and crew.

International Collaboration

Global cooperation is essential for effective mitigation of magnetospheric impacts.

- **Data Sharing:** Promoting the exchange of magnetospheric data and forecasts among countries and organizations.
- **Joint Research Initiatives:** Collaborating on research projects to improve understanding and mitigation of magnetospheric impacts.

4.8 Future Research Directions

Advancements in Magnetospheric Research

Ongoing research aims to enhance our understanding of the magnetosphere and its interactions with the solar wind.

- **New Missions:** Upcoming missions, such as ESA's SMILE (Solar wind Magnetosphere Ionosphere Link Explorer), will provide new insights into the coupling between the solar wind, magnetosphere, and ionosphere.
- **High-Resolution Observations:** Developing instruments with higher resolution and sensitivity to capture detailed observations of magnetospheric phenomena.

Improved Forecasting Techniques

Advances in forecasting techniques will improve our ability to predict magnetospheric activity and mitigate its impacts.

- **Machine Learning and AI:** Using artificial intelligence to analyze magnetospheric data and improve prediction accuracy.
- **Integrated Models:** Developing comprehensive models that integrate solar, magnetospheric, and ionospheric data for more accurate forecasts.

Understanding Long-Term Trends

Studying long-term trends in solar activity and magnetospheric dynamics is essential for predicting future conditions and planning mitigation strategies.

- **Solar Cycle Variability:** Investigating the causes of variability in the solar cycle and its impacts on magnetospheric activity.
- **Historical Data Analysis:** Analyzing historical data from past solar cycles to identify patterns and improve forecasting models.

4.9 Case Studies of Magnetospheric Impacts

The Carrington Event (1859)

One of the most significant geomagnetic storms on record, the Carrington Event, caused widespread disruptions in telegraph systems and produced vivid auroras visible at low latitudes.

- **Cause:** A powerful coronal mass ejection (CME) struck Earth, inducing strong geomagnetic storms.
- **Impact:** Telegraph systems experienced surges and outages, and auroras were seen as far south as the Caribbean.
- **Scientific Insights:** The Carrington Event highlighted the potential for severe space weather impacts and the need for understanding geomagnetic storms.

The March 1989 Geomagnetic Storm

This event caused a major blackout in Quebec, Canada, affecting millions of people and highlighting the vulnerability of power grids to space weather.

- **Cause:** A CME resulted in intense geomagnetic storms and geomagnetically induced currents (GICs).
- **Impact:** The Hydro-Québec power grid collapsed, leaving 6 million people without power for nine hours.
- **Scientific Insights:** This storm emphasized the need for robust power grid protection measures and real-time monitoring of space weather conditions.

The Halloween Storms (2003)

A series of intense solar storms in October-November 2003 disrupted satellite operations, communication systems, and power grids worldwide.

- **Cause:** Multiple CMEs and solar flares led to severe geomagnetic activity.
- **Impact:** Satellite anomalies, aviation disruptions, and power grid issues were reported globally.
- **Scientific Insights:** The Halloween Storms underscored the importance of international collaboration in space weather monitoring and mitigation efforts.

4.10 Technological Advancements and Space Weather Resilience

Next-Generation Satellites

Developing next-generation satellites with enhanced resilience to space weather is critical for ensuring the continuity of space-based services.

- **Radiation-Hardened Electronics:** Incorporating radiation-hardened components to withstand high-energy particle radiation.
- **Real-Time Monitoring:** Equipping satellites with instruments to monitor space weather conditions in real-time and adjust operations accordingly.

Smart Grid Technology

Implementing smart grid technology can improve the resilience of power grids to geomagnetic storms.

- **GIC Blocking Devices:** Installing devices to block or divert geomagnetically induced currents.
- **Adaptive Grid Management:** Using real-time data and AI to dynamically manage power distribution and mitigate the effects of geomagnetic disturbances.

Aviation Safety Measures

Enhancing aviation safety measures to protect passengers and crew from increased radiation exposure during solar events is crucial.

- **Flight Path Adjustments:** Implementing procedures to reroute flights away from polar regions during solar storms.

- **Radiation Monitoring:** Equipping aircraft with radiation monitoring instruments to assess exposure levels and take appropriate action.

4.11 International Policy and Cooperation

Global Space Weather Programs

International cooperation is essential for effective space weather monitoring and mitigation.

- **International Space Weather Initiative (ISWI):** A global program that promotes space weather research, education, and data sharing among participating countries.
- **World Meteorological Organization (WMO):** The WMO collaborates with space weather agencies to integrate space weather forecasting into global meteorological services.

Standardizing Space Weather Data

Developing international standards for space weather data collection, analysis, and dissemination can enhance global preparedness.

- **ISO Standards:** Implementing ISO standards for space weather data to ensure consistency and reliability across different regions and organizations.
- **Global Data Sharing Platforms:** Establishing platforms for sharing real-time space weather data and forecasts among countries and institutions.

4.12 Educational and Public Awareness Initiatives

Space Weather Education Programs

Educational programs can raise awareness about space weather and its impacts on technology and daily life.

- **School Curricula:** Integrating space weather topics into school curricula to educate students about solar activity and its effects.
- **University Courses:** Offering specialized courses in space weather and space physics at universities to train the next generation of scientists and engineers.

Public Outreach and Engagement

Public outreach initiatives can inform the general public about space weather and encourage preparedness.

- **Workshops and Seminars:** Hosting workshops and seminars to educate communities and industry stakeholders about space weather impacts and mitigation strategies.
- **Media Campaigns:** Using media campaigns to raise awareness about space weather and promote protective measures.

The Earth's magnetosphere is a critical shield that protects our planet from the harmful effects of the solar wind and cosmic rays. Understanding its structure, dynamics, and interactions with the solar wind is essential for safeguarding our technological infrastructure and ensuring the safety of human activities in space and on Earth.

This chapter has provided a comprehensive overview of the magnetosphere, including its structure, dynamic processes, and the impacts of magnetospheric disturbances on technology and human activities. By exploring case studies of significant space weather events and discussing advancements in technology and forecasting, we have highlighted the importance of preparedness and resilience in the face of space weather challenges.

As we look to the future, continued research, international cooperation, and public awareness will be essential for addressing the complexities of space weather and ensuring the safety and reliability of our technological infrastructure. By investing in education, innovation, and global collaboration, we can better prepare for and mitigate the impacts of magnetospheric disturbances and other space weather phenomena.

1. **Magnetosphere Structure and Dynamics:** Understanding the Earth's magnetosphere and its interactions with the solar wind is crucial for predicting space weather impacts.
2. **Geomagnetic Storms and Substorms:** These disturbances can have significant effects on technology and human activities, highlighting the need for robust mitigation strategies.
3. **Technological and Operational Mitigation:** Developing resilient technologies and adjusting operational procedures can help minimize the risks posed by space weather.

4. **International Collaboration:** Global cooperation in space weather monitoring, research, and data sharing is essential for effective mitigation of space weather impacts.
5. **Educational and Public Awareness:** Raising awareness about space weather and its effects is crucial for fostering a resilient society.

This comprehensive chapter on the Earth's magnetosphere sets the stage for understanding the subsequent chapters, which will explore specific space weather phenomena, their impacts on technology, and strategies for mitigation and preparedness.

CHAPTER 5: RADIATION BELTS AND THEIR DYNAMICS

The Earth's radiation belts, commonly known as the Van Allen belts, are regions of energetic charged particles trapped by the planet's magnetic field. These belts play a significant role in space weather phenomena and have critical implications for interplanetary travel and human missions. This chapter provides an in-depth analysis of the structure, dynamics, and impacts of the Earth's radiation belts, along with a discussion on the radiation environments within the heliosphere.

5.1 Introduction to Radiation Belts

Discovery and Historical Context

The Van Allen radiation belts were discovered in 1958 by the Explorer 1 and Explorer 3 missions, led by Dr. James Van Allen. These discoveries marked a significant milestone in space science, revealing the presence of high-energy particles trapped by Earth's magnetic field.

Structure and Composition

The radiation belts are composed of two main regions:

- **Inner Radiation Belt:** Extends from about 1 to 3 Earth radii (RE) above the equator. It primarily consists of high-energy protons (tens of MeV) and is relatively stable.
- **Outer Radiation Belt:** Extends from about 3 to 10 RE and contains a mix of electrons and protons. This belt is highly dynamic and influenced by solar activity.

Key Parameters

- **Energy Range:** Particles in the radiation belts have energies ranging from a few keV to several MeV.
- **Particle Flux:** The flux of energetic particles varies with geomagnetic activity and solar wind conditions.
- **Spatial Distribution:** The belts are not uniformly distributed; their shape and intensity are influenced by Earth's magnetic field and solar wind interactions.

5.2 Dynamics of the Radiation Belts

Sources and Acceleration Mechanisms

- **Cosmic Rays:** Galactic cosmic rays (GCRs) and solar energetic particles (SEPs) contribute to the population of the radiation belts.
- **Wave-Particle Interactions:** Interactions with plasma waves, such as chorus, hiss, and magnetosonic waves, can accelerate particles to high energies.

Loss Mechanisms

- **Atmospheric Drag:** Low-energy particles can be lost through collisions with atmospheric particles, leading to atmospheric drag.
- **Magnetopause Shadowing:** High-energy particles can escape through the magnetopause due to dynamic processes in the magnetosphere.
- **Wave-Particle Interactions:** Interactions with waves can scatter particles into the loss cone, causing them to precipitate into the atmosphere.

Temporal Variations

- **Solar Cycle Effects:** The radiation belts exhibit variations linked to the 11-year solar cycle, with increased activity during solar maximum.
- **Geomagnetic Storms:** Intense geomagnetic storms can lead to rapid changes in the belts' structure and particle population.

5.3 Interaction with the Solar Wind

Solar Wind Dynamics

The solar wind, a stream of charged particles emanating from the Sun, interacts with Earth's magnetosphere, influencing the radiation belts.

- **Magnetic Reconnection:** This process allows solar wind energy to enter the magnetosphere, driving geomagnetic storms and substorms.
- **Pressure Changes:** Variations in solar wind pressure can compress or expand the magnetosphere, affecting the radiation belts.

Impact of Coronal Mass Ejections (CMEs)

CMEs, massive bursts of solar wind and magnetic fields, can significantly perturb the radiation belts.

- **Injection of Energetic Particles:** CMEs can inject large numbers of energetic particles into the radiation belts, increasing their intensity.
- **Magnetic Field Reconfigurations:** CMEs can cause major reconfigurations of Earth's magnetic field, altering the structure of the belts.

5.4 Implications for Space Weather and Technological Systems

Satellite Operations

The radiation belts pose significant risks to satellites and space-based technologies.

- **Radiation Damage:** Energetic particles can cause damage to satellite electronics, leading to malfunctions or failures.
- **Surface Charging:** Differential charging of satellite surfaces can lead to electrostatic discharges, damaging components.

Human Spaceflight

Astronauts are at risk of increased radiation exposure when passing through the radiation belts.

- **Health Risks:** Prolonged exposure to high-energy particles can increase the risk of cancer and other health issues.
- **Protective Measures:** Shielding and operational procedures are essential to protect astronauts during missions that traverse the radiation belts.

Ground-Based Impacts

Radiation belt dynamics can also impact ground-based technologies.

- **Communication Disruptions:** High-energy particles can disrupt HF radio communications and GPS signals.
- **Power Grid Vulnerability:** Geomagnetically induced currents (GICs) can be triggered by changes in the radiation belts, affecting power grids.

5.5 Heliospheric Radiation Environments

The Heliosphere

The heliosphere is the vast bubble-like region of space dominated by the solar wind, extending well beyond the orbit of Pluto.

- **Structure:** The heliosphere consists of several regions, including the heliosheath and the heliopause.
- **Cosmic Ray Modulation:** The solar wind modulates the influx of galactic cosmic rays into the inner solar system.

Radiation Environment in the Heliosphere

The radiation environment within the heliosphere is influenced by solar activity and the solar wind.

- **Galactic Cosmic Rays (GCRs):** High-energy particles originating from outside the solar system, modulated by solar activity.
- **Solar Energetic Particles (SEPs):** High-energy particles ejected during solar flares and CMEs.

Implications for Interplanetary Travel

Understanding the radiation environment in the heliosphere is crucial for the safety of interplanetary missions.

- **Radiation Exposure:** Long-duration missions, such as those to Mars, require careful consideration of radiation exposure and protective measures.
- **Shielding Strategies:** Developing effective shielding materials and strategies is essential to protect astronauts from high-energy particles.

5.6 Monitoring and Forecasting Radiation Belts

Space-Based Observatories

Numerous spacecraft monitor the radiation belts and provide valuable data on their dynamics.

- **Van Allen Probes:** These probes provided detailed observations of the radiation belts, helping to improve our understanding of their behavior.
- **GOES Satellites:** These satellites monitor space weather and provide real-time data on radiation belt conditions.

Ground-Based Observatories

Ground-based instruments complement space-based observations by monitoring geomagnetic activity and ionospheric conditions.

- **Magnetometers:** Measure variations in Earth's magnetic field, providing data on geomagnetic storms and substorms.
- **Ionosondes:** Measure ionospheric properties, helping to assess the impact of radiation belt dynamics on communications.

Forecasting Models

Advanced modeling techniques are essential for predicting changes in the radiation belts.

- **MHD Simulations:** Magnetohydrodynamic models simulate the behavior of the radiation belts and their response to solar wind conditions.
- **Empirical Models:** Based on historical data, these models predict the likelihood and severity of space weather events affecting the radiation belts.

5.7 Mitigation Strategies

Technological Solutions

Developing resilient technologies is crucial for mitigating the impacts of radiation belt dynamics.

- **Radiation-Hardened Electronics:** Incorporating radiation-hardened components to withstand high-energy particle radiation.
- **Shielding Materials:** Developing advanced materials to shield satellites and spacecraft from energetic particles.

Operational Strategies

Adjusting operational procedures can help minimize the risks posed by radiation belt dynamics.

- **Satellite Operations:** Adjusting satellite orbits and operations during periods of high radiation belt activity to minimize exposure.
- **Mission Planning:** Timing missions to avoid periods of intense geomagnetic activity and optimizing flight paths to reduce radiation exposure.

International Collaboration

Global cooperation is essential for effective mitigation of radiation belt impacts.

- **Data Sharing:** Promoting the exchange of radiation belt data and forecasts among countries and organizations.
- **Joint Research Initiatives:** Collaborating on research projects to improve understanding and mitigation of radiation belt dynamics.

5.8 Future Research Directions

Advancements in Radiation Belt Research

Ongoing research aims to enhance our understanding of the radiation belts and their interactions with the solar wind.

- **New Missions:** Upcoming missions, such as NASA's Radiation Belt Storm Probes (RBSP), will provide new insights into the dynamics of the radiation belts.
- **High-Resolution Observations:** Developing instruments with higher resolution and sensitivity to capture detailed observations of radiation belt phenomena.

Improved Forecasting Techniques

Advances in forecasting techniques will improve our ability to predict radiation belt dynamics and mitigate their impacts.

- **Machine Learning and AI:** Using artificial intelligence to analyze radiation belt data and improve prediction accuracy.
- **Integrated Models:** Developing comprehensive models that integrate solar, magnetospheric, and ionospheric data for more accurate forecasts.

Understanding Long-Term Trends

Studying long-term trends in solar activity and radiation belt dynamics is essential for predicting future conditions and planning mitigation strategies.

- **Solar Cycle Variability:** Investigating the causes of variability in the solar cycle and its impacts on radiation belt activity.

- **Historical Data Analysis:** Analyzing historical data from past solar cycles to identify patterns and improve forecasting models.

5.9 Case Studies of Radiation Belt Impacts

The South Atlantic Anomaly

The South Atlantic Anomaly (SAA) is a region where Earth's inner radiation belt comes closest to the surface, resulting in increased radiation exposure.

- **Cause:** The SAA is caused by the tilt and offset of Earth's magnetic field, bringing the inner belt closer to the surface.
- **Impact:** Satellites and spacecraft passing through the SAA experience increased radiation exposure, requiring enhanced shielding and operational adjustments.

The Halloween Storms (2003)

The Halloween Storms provide a case study of how intense solar activity can dramatically affect the radiation belts.

- **Cause:** A series of CMEs and solar flares led to significant perturbations in the radiation belts.
- **Impact:** The storms caused widespread disruptions in satellite operations, communication systems, and power grids. The increase in high-energy particles within the radiation belts significantly enhanced radiation exposure, necessitating protective measures for astronauts on the International Space Station (ISS) and other space missions.

The March 1991 Storm

Another significant event was the March 1991 geomagnetic storm, which led to the creation of a new radiation belt.

- **Cause:** A powerful CME struck Earth's magnetosphere, injecting a large number of energetic particles.
- **Impact:** This event resulted in the formation of a transient third radiation belt, demonstrating the dynamic and variable nature of the radiation belts. This temporary belt persisted for several weeks before dissipating.

The Van Allen Probes Discoveries (2012-2019)

The Van Allen Probes, launched in 2012, have provided unprecedented insights into the radiation belts' behavior.

- **Findings**: The probes discovered a temporary third radiation belt, challenging previous models of the belts' structure and dynamics. They also provided detailed data on the processes that accelerate and scatter particles within the belts.
- **Impact**: These findings have significant implications for space weather forecasting and the design of satellite shielding.

5.10 Technological Advancements and Space Weather Resilience

Next-Generation Satellites

Developing next-generation satellites with enhanced resilience to space weather is critical for ensuring the continuity of space-based services.

- **Radiation-Hardened Electronics**: Incorporating radiation-hardened components to withstand high-energy particle radiation. Techniques such as using insulating substrates and hardened transistor designs are common strategies.
- **Real-Time Monitoring**: Equipping satellites with instruments to monitor space weather conditions in real-time and adjust operations accordingly. Instruments like dosimeters and magnetometers help in assessing the radiation environment.

Smart Grid Technology

Implementing smart grid technology can improve the resilience of power grids to geomagnetic storms.

- **GIC Blocking Devices**: Installing devices to block or divert geomagnetically induced currents, thus protecting transformers and other critical components.
- **Adaptive Grid Management**: Using real-time data and AI to dynamically manage power distribution and mitigate the effects of geomagnetic disturbances.

Aviation Safety Measures

Enhancing aviation safety measures to protect passengers and crew from increased radiation exposure during solar events is crucial.

- **Flight Path Adjustments:** Implementing procedures to reroute flights away from polar regions during solar storms to reduce radiation exposure.
- **Radiation Monitoring:** Equipping aircraft with radiation monitoring instruments to assess exposure levels and take appropriate action. Airlines can use dosimeters to measure radiation doses received during flights.

5.11 International Policy and Cooperation

Global Space Weather Programs

International cooperation is essential for effective space weather monitoring and mitigation.

- **International Space Weather Initiative (ISWI):** A global program that promotes space weather research, education, and data sharing among participating countries.
- **World Meteorological Organization (WMO):** The WMO collaborates with space weather agencies to integrate space weather forecasting into global meteorological services.

Standardizing Space Weather Data

Developing international standards for space weather data collection, analysis, and dissemination can enhance global preparedness.

- **ISO Standards:** Implementing ISO standards for space weather data to ensure consistency and reliability across different regions and organizations.
- **Global Data Sharing Platforms:** Establishing platforms for sharing real-time space weather data and forecasts among countries and institutions.

5.12 Educational and Public Awareness Initiatives

Space Weather Education Programs

Educational programs can raise awareness about space weather and its impacts on technology and daily life.

- **School Curricula**: Integrating space weather topics into school curricula to educate students about solar activity and its effects.
- **University Courses**: Offering specialized courses in space weather and space physics at universities to train the next generation of scientists and engineers.

Public Outreach and Engagement

Public outreach initiatives can inform the general public about space weather and encourage preparedness.

- **Workshops and Seminars**: Hosting workshops and seminars to educate communities and industry stakeholders about space weather impacts and mitigation strategies.
- **Media Campaigns**: Using media campaigns to raise awareness about space weather and promote protective measures.

The Earth's radiation belts are dynamic and complex regions of space that play a critical role in space weather phenomena. Understanding their structure, dynamics, and interactions with the solar wind is essential for safeguarding our technological infrastructure and ensuring the safety of human activities in space and on Earth.

This chapter has provided a comprehensive overview of the radiation belts, including their discovery, structure, dynamics, and impacts on technology and human activities. By exploring case studies of significant space weather events and discussing advancements in technology and forecasting, we have highlighted the importance of preparedness and resilience in the face of space weather challenges.

As we look to the future, continued research, international cooperation, and public awareness will be essential for addressing the complexities of space weather and ensuring the safety and reliability of our technological infrastructure. By investing in education, innovation, and global collaboration, we can better prepare for and mitigate the impacts of radiation belt dynamics and other space weather phenomena.

1. **Radiation Belt Structure and Dynamics**: Understanding the Earth's radiation belts and their interactions with the solar wind is crucial for predicting space weather impacts.

2. **Geomagnetic Storms and Substorms:** These disturbances can have significant effects on technology and human activities, highlighting the need for robust mitigation strategies.
3. **Technological and Operational Mitigation:** Developing resilient technologies and adjusting operational procedures can help minimize the risks posed by radiation belt dynamics.
4. **International Collaboration:** Global cooperation in space weather monitoring, research, and data sharing is essential for effective mitigation of space weather impacts.
5. **Educational and Public Awareness:** Raising awareness about space weather and its effects is crucial for fostering a resilient society.

This comprehensive chapter on the Earth's radiation belts sets the stage for understanding the subsequent chapters, which will explore specific space weather phenomena, their impacts on technology, and strategies for mitigation and preparedness.

The ionosphere is a crucial layer of Earth's atmosphere that plays a significant role in communication and navigation systems, especially in the context of space weather. This chapter delves into the structure and behavior of the ionosphere, the effects of space weather on communication and navigation systems, and the broader impacts of space weather on technology.

6.1 Structure and Behavior of the Ionosphere

Introduction to the Ionosphere

The ionosphere extends from about 60 km to over 1,000 km above Earth's surface. It consists of regions with high concentrations of ions and free electrons, created by the ionization of atmospheric gases due to solar and cosmic radiation.

- **D Layer:** Extending from 60 to 90 km, this layer primarily absorbs high-frequency (HF) radio waves and is most prominent during the daytime.
- **E Layer:** Located between 90 and 120 km, the E layer reflects HF radio waves and facilitates long-distance radio communication.
- **F Layer:** The highest layer, extending from 120 km to over 1,000 km, is divided into F1 and F2 layers during the day. The F2 layer is critical for HF radio propagation.

The ionosphere is dynamic, showing significant variations with the time of day, seasons, solar activity, and geomagnetic conditions. During the day, solar radiation ionizes atmospheric particles, increasing the density of ions and free electrons. At night, the absence of solar radiation allows recombination to dominate, reducing ionization levels (Center for Science Education, 2023).

6.2 Effects of Space Weather on the Ionosphere

Solar Radiation and Flares

Solar flares are intense bursts of radiation from the Sun that can cause sudden increases in ionization in the ionosphere. These events can lead to sudden ionospheric disturbances (SIDs), resulting in HF radio blackouts.

Coronal Mass Ejections (CMEs)

CMEs are large expulsions of plasma and magnetic fields from the Sun's corona. When these interact with Earth's magnetosphere, they can cause geomagnetic storms, leading to significant changes in ionospheric density and structure. This can disrupt communication and navigation systems (NASA, 2023).

Solar Wind and Magnetospheric Interaction

The solar wind's interaction with Earth's magnetosphere can enhance auroral activity, which increases ionization in the polar regions. This interaction can cause fluctuations in the ionosphere, affecting radio communications and satellite operations.

Galactic Cosmic Rays (GCRs)

GCRs, high-energy particles from outside the solar system, contribute to ionization in the ionosphere, especially during periods of low solar activity when the Sun's magnetic field is weaker and allows more GCRs to penetrate the solar system.

6.3 Space Weather Effects on Communication Systems

High-Frequency (HF) Radio Communications

HF radio communications rely on the ionosphere for signal reflection. Solar flares and other space weather phenomena can cause SIDs, leading to HF radio blackouts and degraded communication quality.

Very High Frequency (VHF) and Ultra High Frequency (UHF) Communications

Ionospheric scintillation, caused by irregularities in the ionosphere, can affect VHF and UHF communications. This can result in signal fading and loss, impacting satellite communication and broadcasting (Radio JOVE, 2023).

Satellite Communication Systems

Satellites used for communication and navigation are particularly vulnerable to space weather effects. Increased ionization can cause signal delays and degradation, affecting GPS accuracy and reliability. Scintillation can lead to rapid fluctuations in signal strength, disrupting satellite communications (Springer, 2023).

6.4 Space Weather Effects on Navigation Systems

Global Navigation Satellite Systems (GNSS)

GNSS, including GPS, rely on accurate signal transmission through the ionosphere. Ionospheric disturbances can cause delays and signal bending, leading to errors in positioning and timing. Scintillation can degrade the accuracy of GNSS services, impacting navigation for aviation, maritime, and land-based applications.

Aviation Navigation

Aviation relies heavily on GNSS for navigation. Ionospheric disturbances can cause inaccuracies, requiring route adjustments to ensure safety. During intense space weather events, high-altitude flights may experience increased radiation exposure, necessitating precautionary measures.

Maritime Navigation

Maritime navigation systems, which depend on GNSS, can also be affected by ionospheric disturbances. Errors in positioning due to ionospheric delays and scintillation can impact the safety and efficiency of maritime operations.

6.5 Broader Impacts of Space Weather on Technology

Power Grids

Geomagnetic storms can induce geomagnetically induced currents (GICs) in power grids, leading to disruptions and damage. GICs can cause overheating and damage to transformers, resulting in power outages and voltage instability.

Oil and Gas Pipelines

Pipelines are susceptible to GICs, which can accelerate corrosion and cause operational issues. Increased maintenance and operational disruptions can result from GIC-induced interference with pipeline control systems.

Satellite Operations

Space weather can affect satellite operations by causing radiation damage to electronics, differential charging leading to electrostatic discharges, and increased atmospheric drag during geomagnetic storms, which can result in orbit decay and re-entry.

6.6 Monitoring and Forecasting Ionospheric Disturbances

Space-Based Observatories

Space missions like NASA's ICON (Ionospheric Connection Explorer) and ESA's Swarm provide valuable data on the ionosphere's response to space weather, helping to improve our understanding and forecasting capabilities.

Ground-Based Observatories

Ground-based instruments, such as magnetometers and ionosondes, monitor geomagnetic activity and ionospheric conditions. These observations are crucial for real-time assessment and forecasting of ionospheric disturbances (Springer, 2023).

Data Analysis and Modeling

Advanced data analysis and modeling techniques, including machine learning and numerical simulations, are essential for predicting ionospheric disturbances and their impacts on communication and navigation systems. Integrated models that combine solar, magnetospheric, and ionospheric data can provide more accurate forecasts.

6.7 Mitigation Strategies for Ionospheric Disturbances

Technological Solutions

Developing resilient technologies is crucial for mitigating the impacts of ionospheric disturbances. This includes radiation-hardened electronics for communication and navigation systems and adaptive systems that can dynamically adjust to changing ionospheric conditions.

Operational Strategies

Operational adjustments, such as frequency management and timing adjustments, can help minimize the risks posed by ionospheric disturbances. Using multiple frequencies and dynamic frequency selection can avoid affected frequencies, while timing adjustments can account for ionospheric delays.

International Collaboration

Global cooperation is essential for effective mitigation of ionospheric disturbances. Data sharing and joint research initiatives, such as those facilitated by the International GNSS Service (IGS) and the International Space Weather Initiative (ISWI), promote a coordinated approach to space weather monitoring and research.

6.8 Future Research Directions

Advancements in Ionospheric Research

Future research aims to enhance our understanding of the ionosphere and its response to space weather. Missions like NASA's GOLD (Global-scale Observations of the Limb and Disk) will provide new insights into the ionosphere's behavior and its interactions with the thermosphere.

Improved Forecasting Techniques

Advances in forecasting techniques, including the use of artificial intelligence, will improve our ability to predict ionospheric disturbances and mitigate their impacts. Integrated models that combine solar, magnetospheric, and ionospheric data will provide more accurate forecasts.

Understanding Long-Term Trends

Studying long-term trends in solar activity and ionospheric behavior is essential for predicting future conditions and planning mitigation strategies. Understanding the variability of the solar cycle and its impact on the ionosphere will help improve space weather forecasting.

6.9 Case Studies of Ionospheric Disturbances

The March 1989 Geomagnetic Storm

This significant geomagnetic storm caused a major blackout in Quebec, Canada, and disrupted HF radio communications and GNSS services. It highlighted the vulnerability of technology to ionospheric disturbances.

The Halloween Storms (2003)

These storms caused widespread disruptions in satellite communications, GPS accuracy, and HF radio propagation. The ionospheric disturbances were particularly severe in the polar regions, affecting aviation and maritime navigation.

The St. Patrick's Day Storm (2015)

This storm led to significant ionospheric disturbances, disrupting GNSS services and HF communications. It underscored the need for improved forecasting and mitigation strategies for space weather impacts on modern technology.

The ionosphere is a dynamic region of Earth's atmosphere that plays a crucial role in communication and navigation systems. Understanding its structure, behavior, and response to space weather phenomena is essential for mitigating the impacts of ionospheric disturbances on technology and human activities.

This chapter has provided a comprehensive overview of the ionosphere, including its structure, the effects of space weather on its behavior, and the broader impacts on communication and navigation systems. By exploring case studies of significant ionospheric disturbances and discussing advancements in technology and forecasting, we have highlighted the importance of preparedness and resilience in the face of space weather challenges.

As we look to the future, continued research, international cooperation, and public awareness will be essential for addressing the complexities of space weather and ensuring the safety and reliability of our technological infrastructure. By investing in education, innovation, and global collaboration, we can better prepare for and mitigate the impacts of ionospheric disturbances and other space weather phenomena.

Key Takeaways

1. **Ionosphere Structure and Behavior:** Understanding the ionosphere's structure and behavior is crucial for predicting and mitigating the impacts of space weather on communication and navigation systems.
2. **Space Weather Effects:** Solar flares, CMEs, and other space weather phenomena can cause significant ionospheric disturbances, affecting communication and navigation systems.

3. **Technological and Operational Mitigation:** Developing resilient technologies and adjusting operational procedures can help minimize the risks posed by ionospheric disturbances.
4. **International Collaboration:** Global cooperation in space weather monitoring, research, and data sharing is essential for effective mitigation of space weather impacts.
5. **Educational and Public Awareness:** Raising awareness about space weather and its effects is crucial for fostering a resilient society.

This comprehensive chapter on ionospheric disturbances sets the stage for understanding the subsequent chapters, which will explore specific space weather phenomena, their impacts on technology, and strategies for mitigation and preparedness.

6.10 Mitigation Strategies for Ionospheric Disturbances (continued)

Improving Technological Resilience

Developing resilient technologies is critical for mitigating the impacts of ionospheric disturbances on communication and navigation systems. This includes:

- **Radiation-Hardened Electronics:** Incorporating radiation-hardened components in communication and navigation systems to withstand space weather effects.
- **Adaptive Systems:** Designing systems that can dynamically adjust to changing ionospheric conditions, such as adaptive antennas and frequency-hopping techniques.

Operational Adjustments

Operational adjustments can help minimize the risks posed by ionospheric disturbances. These strategies include:

- **Frequency Management:** Using multiple frequencies and dynamic frequency selection to avoid affected frequencies during ionospheric disturbances.
- **Timing Adjustments:** Adjusting the timing of communication and navigation signals to account for ionospheric delays.

International Collaboration

Global cooperation is essential for effective mitigation of ionospheric disturbances. This involves:

- **Data Sharing:** Promoting the exchange of ionospheric data and forecasts among countries and organizations.
- **Joint Research Initiatives:** Collaborating on research projects to improve understanding and mitigation of ionospheric disturbances. Programs like the Global Ionospheric Radio Observatory (GIRO) and the International Space Weather Initiative (ISWI) encourage international research partnerships.

Education and Public Awareness

Raising awareness about space weather and its effects on the ionosphere is crucial for fostering a resilient society. Educational programs and public outreach initiatives can inform the general public and industry stakeholders about space weather impacts and encourage preparedness.

The ionosphere is a dynamic region of Earth's atmosphere that plays a crucial role in communication and navigation systems. Understanding its structure, behavior, and response to space weather phenomena is essential for mitigating the impacts of ionospheric disturbances on technology and human activities.

This chapter has provided a comprehensive overview of the ionosphere, including its structure, the effects of space weather on its behavior, and the broader impacts on communication and navigation systems. By exploring case studies of significant ionospheric disturbances and discussing advancements in technology and forecasting, we have highlighted the importance of preparedness and resilience in the face of space weather challenges.

As we look to the future, continued research, international cooperation, and public awareness will be essential for addressing the complexities of space weather and ensuring the safety and reliability of our technological infrastructure. By investing in education, innovation, and global collaboration, we can better prepare for and mitigate the impacts of ionospheric disturbances and other space weather phenomena.

This comprehensive chapter on ionospheric disturbances sets the stage for understanding the subsequent chapters, which will explore specific space weather phenomena, their impacts on technology, and strategies for mitigation and preparedness.

CHAPTER 7: SOLAR FLARES AND CORONAL MASS EJECTIONS

Solar flares and coronal mass ejections (CMEs) are two of the most powerful and dynamic phenomena originating from the Sun. These events are critical drivers of space weather and have significant implications for Earth's technological systems and human activities. This chapter explores the mechanisms, characteristics, impacts, and current research on solar flares and CMEs, providing a comprehensive understanding of these solar phenomena.

7.1 Introduction to Solar Flares and Coronal Mass Ejections

Definitions and Distinctions

- **Solar Flares** are intense bursts of radiation resulting from the release of magnetic energy in the Sun's atmosphere. They are observed across the electromagnetic spectrum, from radio waves to X-rays and gamma rays.
- **Coronal Mass Ejections (CMEs)** involve massive bursts of solar wind and magnetic fields rising above the solar corona or being released into space. Unlike flares, CMEs involve the expulsion of billions of tons of plasma and can travel at speeds ranging from 250 to 3000 kilometers per second.

Both phenomena are closely related but distinct in their nature and effects. Solar flares primarily impact Earth's ionosphere, causing radio blackouts, while CMEs can lead to geomagnetic storms by interacting with Earth's magnetosphere.

7.2 Mechanisms and Causes

Solar Flares

Solar flares occur due to magnetic reconnection, a process where twisted magnetic field lines in the Sun's atmosphere suddenly realign, releasing vast amounts of energy. These flares are typically associated with active regions on the Sun, particularly around sunspots where magnetic fields are strongest and most complex.

Coronal Mass Ejections

CMEs are caused by the release of magnetic stress in the solar corona. This stress is often due to the twisting and shearing of magnetic field lines. The energy stored in these magnetic fields is suddenly released, propelling plasma into space. CMEs often originate from active regions, but they can also occur in quieter areas of the Sun where magnetic flux ropes become destabilized.

7.3 Characteristics and Classification

Solar Flares

Solar flares are classified based on their X-ray brightness, measured by the Geostationary Operational Environmental Satellites (GOES):

- **Class A, B, C:** Weak flares with minimal impact.
- **Class M:** Medium-sized flares causing brief radio blackouts.
- **Class X:** The largest flares, capable of causing extensive radio blackouts and radiation storms.

Flares can last from minutes to hours, with their peak intensity often reached within a few minutes of onset.

Coronal Mass Ejections

CMEs vary significantly in size, speed, and structure:

- **Speed:** CMEs can travel at speeds from 250 km/s to nearly 3000 km/s. Fast CMEs can reach Earth in as little as 15-18 hours.
- **Size:** As CMEs travel away from the Sun, they expand and can occupy a large portion of space between the Sun and Earth.
- **Magnetic Structure:** The embedded magnetic field in a CME can interact with Earth's magnetosphere, causing geomagnetic storms. The direction and strength of this magnetic field are crucial for determining the impact on Earth.

7.4 Impacts on Earth and Space

Geospace Environment

- **Geomagnetic Storms:** CMEs interacting with Earth's magnetosphere can induce geomagnetic storms, leading to disturbances in Earth's magnetic field.
- **Auroras:** One visible effect of geomagnetic storms is the enhancement of auroras, visible at lower latitudes during intense storms.

Technological Systems

- **Satellite Operations:** Increased radiation from solar flares and the plasma from CMEs can damage satellites, interfere with their electronics, and disrupt communication and navigation systems.
- **Communication Systems:** HF radio blackouts are common during solar flares, affecting aviation and maritime operations. CMEs can disrupt GNSS signals, leading to navigation errors.
- **Power Grids:** Geomagnetic induced currents (GICs) from geomagnetic storms can damage transformers and other components of power grids, potentially causing widespread power outages.

Human Health

- **Astronaut Safety:** Increased radiation levels from solar flares and CMEs pose significant risks to astronauts. Protective measures, such as sheltering in shielded areas, are essential during solar storms.
- **Aviation:** Increased radiation can also affect high-altitude flights, particularly over polar routes, where the Earth's magnetic field offers less protection.

7.5 Monitoring and Predicting Solar Flares and CMEs

Observation Techniques

- **Space-Based Observatories:** Satellites like SOHO, STEREO, and the Solar Dynamics Observatory (SDO) provide continuous monitoring of the Sun, capturing high-resolution images and data on solar activity.
- **Ground-Based Observatories:** Telescopes and radio observatories on Earth complement space-based observations, offering different perspectives and additional data.

Predictive Models

- **Magnetohydrodynamic (MHD) Models:** These models simulate the behavior of solar plasma and magnetic fields, helping to predict the onset and development of solar flares and CMEs.
- **Empirical Models:** Based on historical data, these models provide probabilistic forecasts of solar activity.
- **Machine Learning:** Advances in machine learning and AI are being integrated into predictive models to improve accuracy and provide real-time alerts.

Early Warning Systems

- **NOAA's Space Weather Prediction Center (SWPC):** Provides real-time alerts and forecasts of solar flares and CMEs, helping to mitigate their impacts on technology and human activities.
- **ESA's Space Weather Coordination Centre (SSCC):** Offers comprehensive space weather monitoring and forecasting services for Europe and beyond.

7.6 Case Studies of Major Solar Flares and CMEs

The Carrington Event (1859)

The Carrington Event remains the most powerful geomagnetic storm on record. It was caused by a massive solar flare and CME. Telegraph systems were severely disrupted, and auroras were seen at low latitudes worldwide.

The Halloween Storms (2003)

A series of intense solar flares and CMEs in late October 2003 caused widespread technological disruptions, including satellite anomalies, power grid disturbances, and communication blackouts globally.

The St. Patrick's Day Storm (2015)

A significant geomagnetic storm triggered by a CME from a solar flare, causing GNSS disruptions, auroras at lower latitudes, and minor power grid issues.

7.7 Mitigation Strategies

Technological Solutions

- **Satellite Shielding:** Enhancing the shielding of satellites to protect against increased radiation levels during solar flares and CMEs.
- **Power Grid Protections:** Implementing measures such as GIC blocking devices and improved transformer designs to mitigate the impacts on power grids.

Operational Procedures

- **Satellite Maneuvers:** Adjusting satellite orbits and orientations to minimize exposure to high-radiation regions.

- **Flight Path Adjustments:** Rerouting flights, particularly over polar regions, during periods of high solar activity to protect passengers and crew from increased radiation.

Public Awareness and Preparedness

- **Education Campaigns:** Informing the public and stakeholders about the risks associated with solar flares and CMEs and the importance of preparedness measures.
- **Early Warning Systems:** Ensuring that early warning systems are in place and effectively communicated to mitigate the impacts on critical infrastructure and daily life.

Solar flares and coronal mass ejections are dynamic and powerful phenomena that significantly impact space weather and modern technological systems. Understanding their mechanisms, characteristics, and effects is crucial for developing effective monitoring, prediction, and mitigation strategies. This chapter has provided a detailed overview of these phenomena, highlighting the importance of continuous research, technological innovation, and international collaboration in addressing the challenges posed by solar flares and CMEs.

1. **Mechanisms and Causes:** Solar flares and CMEs are driven by magnetic reconnection and the release of magnetic energy in the Sun's atmosphere.
2. **Characteristics and Classification:** Solar flares are classified based on their X-ray brightness, while CMEs vary in size, speed, and directionality.
3. **Impacts on Earth and Space:** Both phenomena can cause significant disruptions to satellite operations, communication systems, power grids, and human health.
4. **Monitoring and Predicting:** Advanced observation techniques, predictive models, and early warning systems are essential for mitigating the impacts of solar flares and CMEs.
5. **Mitigation Strategies:** Technological solutions, operational procedures, and public awareness are critical for enhancing resilience against space weather events.

This comprehensive chapter on solar flares and coronal mass ejections provides a solid foundation for understanding their role in space weather and their impacts on Earth and technological systems. Through continued research, technological advancements, and international collaboration, we can enhance our ability to predict and mitigate the effects of these powerful solar events.

Geomagnetic storms are large-scale disturbances in Earth's magnetosphere caused by the efficient transfer of energy from the solar wind into the space environment surrounding Earth. These storms can have significant impacts on various technological systems, including satellite operations, power grids, communication systems, and navigation systems. This chapter delves into the causes, mechanisms, and effects of geomagnetic storms, as well as strategies for monitoring, forecasting, and mitigating their impacts.

8.1 Introduction to Geomagnetic Storms

Definition and Historical Context

A geomagnetic storm is a major disturbance of Earth's magnetosphere that occurs when there is an exchange of energy from the solar wind into the space environment surrounding Earth. These storms are characterized by significant variations in the geomagnetic field and are measured using indices such as the Kp index and the Disturbance Storm Time (Dst) index.

- **Historical Events:** Notable geomagnetic storms include the Carrington Event of 1859, which caused widespread auroras and disruptions in telegraph systems, and the March 1989 storm, which led to a major power outage in Quebec, Canada.

Causes of Geomagnetic Storms

Geomagnetic storms are primarily caused by solar wind disturbances associated with solar activities such as coronal mass ejections (CMEs), high-speed solar wind streams (HSS), and co-rotating interaction regions (CIRs).

- **Coronal Mass Ejections (CMEs):** Large expulsions of plasma and magnetic fields from the Sun's corona that can cause severe geomagnetic storms when directed toward Earth.
- **High-Speed Solar Wind Streams (HSS):** Streams of fast solar wind originating from coronal holes that interact with the slower solar wind, leading to geomagnetic activity.
- **Co-rotating Interaction Regions (CIRs):** Regions where high-speed solar wind streams overtake slower streams, creating compression regions that can drive geomagnetic storms.

8.2 Mechanisms of Geomagnetic Storms

Solar Wind-Magnetosphere Interaction

The interaction between the solar wind and Earth's magnetosphere is the primary mechanism driving geomagnetic storms. Key processes include:

- **Magnetic Reconnection:** The process where oppositely directed magnetic fields reconnect, allowing solar wind energy to enter the magnetosphere. This typically occurs on the dayside magnetopause and in the magnetotail.
- **Bow Shock and Magnetopause Dynamics:** The bow shock is the region where the solar wind slows down abruptly upon encountering Earth's magnetosphere. The magnetopause is the boundary where the pressure from the solar wind is balanced by Earth's magnetic field. Variations in solar wind pressure can compress or expand the magnetosphere.

Energy Transfer and Storage

Energy from the solar wind is transferred into the magnetosphere and stored in the magnetotail. During geomagnetic storms, this energy is released through magnetic reconnection, driving various dynamic processes in the magnetosphere.

- **Ring Current Formation:** The ring current is formed by energetic ions and electrons drifting around Earth. During geomagnetic storms, the ring current intensifies, leading to a global decrease in the geomagnetic field strength, measured by the Dst index.
- **Auroral Substorms:** Substorms are localized disturbances in the magnetosphere that result in the brightening of auroras. They are caused by the sudden release of energy stored in the magnetotail.

8.3 Impacts of Geomagnetic Storms

Satellite Operations

Geomagnetic storms can have several detrimental effects on satellites, including:

- **Radiation Damage:** Increased radiation levels can damage satellite electronics and solar panels, potentially leading to malfunctions or failures.

- **Surface Charging:** Differential charging of satellite surfaces can cause electrostatic discharges, damaging components and leading to operational anomalies.
- **Orbit Decay:** Increased atmospheric drag during geomagnetic storms can cause satellites to lose altitude and eventually re-enter the atmosphere.

Power Grids

Geomagnetic storms can induce geomagnetically induced currents (GICs) in power grids, causing:

- **Transformer Damage:** GICs can cause overheating and damage to transformers, leading to power outages and voltage instability.
- **Voltage Instability:** Fluctuations in voltage caused by GICs can affect the stability of power grids, potentially leading to widespread blackouts.

Communication Systems

Geomagnetic storms can disrupt various communication systems, including:

- **HF Radio Communications:** Increased ionization in the ionosphere can cause HF radio blackouts and signal degradation.
- **Satellite Communications:** Signal delays and scintillation can affect the quality and reliability of satellite communications.

Navigation Systems

Geomagnetic storms can impact navigation systems, including:

- **GPS Accuracy:** Ionospheric disturbances can cause delays and signal bending, leading to errors in GPS positioning and timing.
- **Aviation Navigation:** Increased ionospheric activity can disrupt navigation systems, requiring route adjustments for safety.

8.4 Monitoring and Forecasting Geomagnetic Storms

Space-Based Observatories

Space-based observatories provide critical data on solar and geomagnetic activity, helping to monitor and forecast geomagnetic storms.

- **SOHO (Solar and Heliospheric Observatory):** Provides continuous observations of the Sun's corona and solar wind.
- **ACE (Advanced Composition Explorer):** Monitors solar wind and cosmic rays near the Lagrange point L1.
- **Parker Solar Probe:** Studies the solar corona and the processes that accelerate the solar wind.

Ground-Based Observatories

Ground-based observatories complement space-based observations by monitoring geomagnetic activity and ionospheric conditions.

- **Magnetometers:** Measure variations in Earth's magnetic field, providing data on geomagnetic activity.
- **Ionosondes:** Measure ionospheric properties, helping to assess the impact of geomagnetic storms on communications.

Data Analysis and Modeling

Advanced data analysis and modeling techniques are essential for understanding and predicting geomagnetic storms.

- **MHD Simulations:** Magnetohydrodynamic models simulate the behavior of the magnetosphere and its response to solar wind conditions.
- **Empirical Models:** Based on historical data, these models predict the likelihood and severity of geomagnetic storms.

8.5 Mitigation Strategies for Geomagnetic Storms

Technological Solutions

Developing resilient technologies is crucial for mitigating the impacts of geomagnetic storms on technological systems.

- **Radiation-Hardened Electronics:** Incorporating radiation-hardened components to withstand high-energy particle radiation.
- **GIC Blocking Devices:** Installing devices to block or divert geomagnetically induced currents, protecting power grid infrastructure.

Operational Strategies

Adjusting operational procedures can help minimize the risks posed by geomagnetic storms.

- **Satellite Operations:** Adjusting satellite orbits and operations during geomagnetic storms to minimize exposure to radiation.
- **Aviation Procedures:** Rerouting high-altitude flights during solar events to reduce radiation exposure for passengers and crew.

International Collaboration

Global cooperation is essential for effective mitigation of geomagnetic storms.

- **Data Sharing:** Promoting the exchange of space weather data and forecasts among countries and organizations.
- **Joint Research Initiatives:** Collaborating on research projects to improve understanding and mitigation of geomagnetic storms.

Education and Public Awareness

Raising awareness about geomagnetic storms and their effects is crucial for fostering a resilient society.

- **Educational Programs:** Implementing educational programs to teach students and the general public about space weather and its impacts.
- **Public Outreach:** Conducting public outreach campaigns to inform communities and industry stakeholders about geomagnetic storms and preparedness measures.

8.6 Case Studies of Geomagnetic Storms

The Carrington Event (1859)

One of the most significant geomagnetic storms on record, the Carrington Event, caused widespread disruptions in telegraph systems and produced vivid auroras visible at low latitudes.

- **Cause:** A powerful CME struck Earth, inducing strong geomagnetic storms.
- **Impact:** Telegraph systems experienced surges and outages, and auroras were seen as far south as the Caribbean.

The March 1989 Geomagnetic Storm

This event caused a major blackout in Quebec, Canada, affecting millions of people and highlighting the vulnerability of power grids to geomagnetic storms.

- **Cause:** A CME resulted in intense geomagnetic storms and geomagnetically induced currents (GICs).
- **Impact:** The Hydro-Québec power grid collapsed, leaving 6 million people without power for nine hours.

The Halloween Storms (2003)

A series of intense solar storms in October-November 2003 disrupted satellite operations, communication systems, and power grids worldwide.

- **Cause:** Multiple CMEs and solar flares led to severe geomagnetic activity.
- **Impact:** Satellite anomalies, aviation disruptions, and power grid issues were reported globally.

8.7 Future Research Directions

Advancements in Geomagnetic Storm Research

Ongoing research aims to enhance our understanding of geomagnetic storms and their interactions with the solar wind. Efforts include deploying new space missions, enhancing observational capabilities, and improving forecasting models.

- **New Missions:** Upcoming missions, such as ESA's Solar Orbiter and NASA's IMAP (Interstellar Mapping and Acceleration Probe), will provide new insights into the solar wind and its effects. These missions aim to study the sources of solar wind and how it propagates through space, contributing to better predictions of geomagnetic storms .
- **High-Resolution Observations:** Developing instruments with higher resolution and sensitivity to capture detailed observations of geomagnetic storm phenomena. These advancements help in understanding the fine-scale structures and dynamics of the magnetosphere during storm events .

Improved Forecasting Techniques

Advances in forecasting techniques will improve our ability to predict geomagnetic storms and mitigate their impacts. Integrating new technologies and methods is crucial for enhancing the accuracy and reliability of space weather forecasts.

- **Machine Learning and AI:** Using artificial intelligence to analyze space weather data and improve prediction accuracy. Machine learning algorithms can identify patterns in large datasets and provide real-time forecasts .
- **Integrated Models:** Developing comprehensive models that integrate solar, magnetospheric, and ionospheric data for more accurate forecasts. These models simulate the complex interactions between different space weather components, providing better insights into storm dynamics .

Understanding Long-Term Trends

Studying long-term trends in solar activity and geomagnetic storm dynamics is essential for predicting future conditions and planning mitigation strategies. Historical analysis and pattern recognition are key aspects of this research.

- **Solar Cycle Variability:** Investigating the causes of variability in the solar cycle and its impacts on geomagnetic storm activity. Understanding these long-term trends helps in predicting periods of increased space weather activity .
- **Historical Data Analysis:** Analyzing historical data from past solar cycles to identify patterns and improve forecasting models. Long-term datasets from observatories and satellite missions are valuable resources for this research .

8.8 Educational and Public Awareness Initiatives

Educational Programs

Educational programs are essential for raising awareness about geomagnetic storms and their impacts. Implementing space weather topics in school curricula and offering specialized courses at universities can foster a deeper understanding of these phenomena.

- **School Curricula:** Integrating space weather topics into science education programs to teach students about solar activity and its effects on Earth.

- **University Courses:** Offering specialized courses in space weather, magnetospheric physics, and related fields to train the next generation of scientists and engineers .

Public Outreach and Engagement

Public outreach initiatives can inform the general public and industry stakeholders about geomagnetic storms and encourage preparedness. Effective communication strategies include workshops, seminars, media campaigns, and the use of digital platforms.

- **Workshops and Seminars:** Hosting events to educate communities and professionals about space weather impacts and mitigation strategies.
- **Media Campaigns:** Utilizing various media channels to raise awareness about space weather phenomena and promote protective measures .

Geomagnetic storms are complex and impactful space weather events driven by the interaction between the solar wind and Earth's magnetosphere. Understanding their causes, mechanisms, and effects is crucial for mitigating the risks they pose to technological systems and human activities.

This chapter has provided a comprehensive overview of geomagnetic storms, including their causes, dynamic processes, and impacts on technology and infrastructure. By exploring case studies of significant geomagnetic storms and discussing advancements in research, forecasting, and mitigation strategies, we have highlighted the importance of preparedness and resilience in the face of these space weather challenges.

Looking ahead, continued research, international collaboration, and public awareness will be essential for addressing the complexities of geomagnetic storms and ensuring the safety and reliability of our technological infrastructure. By investing in education, innovation, and global cooperation, we can better prepare for and mitigate the impacts of geomagnetic storms and other space weather phenomena.

1. **Understanding Geomagnetic Storms:** Comprehensive knowledge of the causes, mechanisms, and impacts of geomagnetic storms is essential for predicting and mitigating their effects.
2. **Impact on Technology:** Geomagnetic storms can significantly affect satellite operations, power grids, communication systems, and navigation systems, highlighting the need for robust mitigation strategies.

3. **Technological and Operational Mitigation:** Developing resilient technologies and adjusting operational procedures can help minimize the risks posed by geomagnetic storms.
4. **Future Research and Forecasting:** Advancements in research, observation, and forecasting techniques are crucial for improving our ability to predict geomagnetic storms and mitigate their impacts.
5. **International Collaboration and Public Awareness:** Global cooperation and effective public outreach are essential for fostering a resilient society capable of addressing space weather challenges.

This comprehensive chapter on geomagnetic storms sets the stage for understanding the subsequent chapters, which will explore specific space weather phenomena, their impacts on technology, and strategies for mitigation and preparedness.

CHAPTER 9: IMPACT ON SATELLITE OPERATIONS

Satellites are integral to modern communication, navigation, weather forecasting, and scientific research. However, these sophisticated devices are vulnerable to space weather events such as solar flares, coronal mass ejections (CMEs), and geomagnetic storms. This chapter explores the various ways space weather impacts satellite operations, the underlying mechanisms, and strategies for mitigation and resilience.

9.1 Overview of Satellite Operations

Types of Satellites and Their Functions

Satellites serve various purposes, including communication, navigation, Earth observation, weather forecasting, and scientific research.

- **Communication Satellites:** Facilitate global telecommunications, including television broadcasting, internet services, and secure military communications.
- **Navigation Satellites:** Provide positioning, navigation, and timing services critical for transportation, logistics, and emergency response (e.g., GPS, GLONASS, Galileo).
- **Earth Observation Satellites:** Monitor environmental conditions, natural disasters, and climate change (e.g., Landsat, Copernicus).
- **Weather Satellites:** Track weather patterns and provide data for forecasting and climate studies (e.g., GOES, Meteosat).
- **Scientific Satellites:** Conduct space research and gather data on astronomical and geophysical phenomena (e.g., Hubble Space Telescope, Parker Solar Probe).

9.2 Vulnerability of Satellites to Space Weather

Radiation Exposure

Satellites are exposed to various forms of radiation from the Sun and cosmic rays, which can damage their electronics and affect their operation.

- **Solar Flares:** Emit high-energy particles and X-rays that can penetrate satellite shielding, causing single-event upsets (SEUs) and damaging electronic components.

- **Cosmic Rays:** High-energy particles from outside the solar system can also cause SEUs and long-term degradation of satellite components.
- **Radiation Belts:** The Van Allen belts pose a significant radiation hazard, especially for satellites in geostationary and medium Earth orbits.

Electrostatic Charging

Space weather events can cause differential charging of satellite surfaces, leading to electrostatic discharges that can damage electronic systems.

- **Surface Charging:** Occurs when different parts of the satellite accumulate different electric potentials, leading to arcing and potential damage.
- **Internal Charging:** Caused by high-energy particles penetrating the satellite's shielding and depositing charge inside, potentially leading to discharges and damage to internal components.

Geomagnetic Storms

Geomagnetic storms can cause significant disturbances in the Earth's magnetosphere, affecting satellite operations in several ways.

- **Increased Atmospheric Drag:** Enhanced ionization and heating of the upper atmosphere during geomagnetic storms increase drag on low Earth orbit (LEO) satellites, potentially leading to orbit decay and re-entry.
- **Magnetospheric Substorms:** Can cause rapid changes in the magnetic environment, inducing currents in satellite structures and affecting their stability and control systems.

9.3 Impacts on Specific Satellite Systems

Communication Satellites

Communication satellites are particularly vulnerable to space weather due to their reliance on stable orbits and clear signal transmission paths.

- **Signal Degradation:** Increased ionization in the ionosphere can cause signal scattering and fading, degrading the quality of communication signals.
- **Transponder Damage:** High-energy particles can damage the satellite's transponders, reducing its ability to relay signals effectively.

Navigation Satellites

Navigation satellites provide critical services for positioning and timing, making them highly sensitive to space weather disruptions.

- **Positioning Errors:** Ionospheric disturbances can cause delays and signal bending, leading to inaccuracies in positioning data.
- **Signal Scintillation:** Rapid fluctuations in signal strength due to ionospheric irregularities can degrade the accuracy and reliability of navigation systems.

Earth Observation Satellites

Earth observation satellites require stable orbits and precise instruments to monitor environmental conditions accurately.

- **Instrument Degradation:** Radiation exposure can degrade the performance of imaging and sensing instruments, reducing data quality.
- **Orbital Decay:** Increased atmospheric drag during geomagnetic storms can affect the satellite's orbit, impacting its ability to capture consistent data.

Weather Satellites

Weather satellites rely on accurate data collection and transmission to monitor and predict weather patterns.

- **Data Transmission Issues:** Signal degradation and delays caused by space weather can affect the timely transmission of weather data.
- **Instrument Calibration:** Radiation exposure can impact the calibration of instruments, affecting the accuracy of weather predictions.

Scientific Satellites

Scientific satellites conduct critical research and gather data on astronomical and geophysical phenomena.

- **Data Integrity:** Radiation and electrostatic discharges can affect the integrity of scientific data, leading to potential gaps or inaccuracies.
- **Operational Interruptions:** Space weather events can cause temporary or permanent disruptions in the satellite's operation, affecting ongoing research missions.

9.4 Monitoring and Mitigation Strategies

Space-Based Observatories

Space-based observatories play a crucial role in monitoring space weather and providing data to protect satellite operations.

- **SOHO (Solar and Heliospheric Observatory):** Monitors the Sun's activity and provides early warnings of solar flares and CMEs.
- **ACE (Advanced Composition Explorer):** Measures solar wind and cosmic rays near the Lagrange point L1, providing real-time data on space weather conditions.
- **Parker Solar Probe:** Studies the solar corona and solar wind to improve understanding and prediction of space weather events.

Ground-Based Observatories

Ground-based instruments complement space-based observations by monitoring geomagnetic activity and ionospheric conditions.

- **Magnetometers:** Measure variations in Earth's magnetic field, providing data on geomagnetic storms and substorms.
- **Ionosondes:** Measure ionospheric properties, helping to assess the impact of space weather on satellite communications and navigation.

Technological Solutions

Developing resilient technologies is essential for protecting satellites from space weather impacts.

- **Radiation-Hardened Electronics:** Incorporating radiation-hardened components to withstand high-energy particle radiation.
- **Shielding:** Using materials and designs that provide better protection against radiation and electrostatic charging.
- **Redundancy and Fault Tolerance:** Designing satellites with redundant systems and fault-tolerant architectures to ensure continued operation during space weather events.

Operational Strategies

Adjusting operational procedures can help mitigate the risks posed by space weather.

- **Orbit Adjustments:** Altering satellite orbits to minimize exposure to radiation belts and other hazardous regions during space weather events.
- **Power Management:** Reducing power loads and shutting down non-essential systems during high-radiation periods to protect sensitive components.
- **Communication Protocols:** Implementing robust communication protocols to handle signal degradation and ensure reliable data transmission.

International Collaboration

Global cooperation is crucial for effective space weather monitoring and mitigation.

- **Data Sharing:** Promoting the exchange of space weather data and forecasts among countries and organizations.
- **Joint Research Initiatives:** Collaborating on research projects to improve understanding and mitigation of space weather impacts on satellite operations.

Public Awareness and Education

Raising awareness about space weather and its impacts on satellite operations is essential for fostering a resilient society.

- **Educational Programs:** Implementing educational programs to teach students and the general public about space weather and its effects on technology.
- **Public Outreach:** Conducting public outreach campaigns to inform communities and industry stakeholders about space weather impacts and preparedness measures.

9.5 Causes and Magnitude of Space Weather Impacts

Solar Flares and CMEs

Solar flares and coronal mass ejections (CMEs) are primary drivers of space weather impacts on satellites. Solar flares release intense bursts of radiation, while CMEs eject large volumes of plasma and magnetic fields into space. When directed toward Earth, these events can cause significant geomagnetic disturbances and radiation storms.

Geomagnetic Storms and Substorms

Geomagnetic storms, triggered by interactions between CMEs and Earth's magnetosphere, can cause large-scale disruptions. Substorms, which are localized disturbances within the magnetosphere, can also have significant impacts on satellite operations.

Radiation Belts

The Van Allen radiation belts, consisting of trapped high-energy particles, pose a constant threat to satellites. Changes in the radiation environment during space weather events can increase the intensity and variability of these belts, exacerbating the risks to satellite electronics.

Magnitude of Impacts

The severity of space weather impacts on satellites can vary widely. Minor events may cause temporary communication disruptions, while major storms can lead to satellite damage, loss of functionality, or even complete failure. For example, in February 2022, a geomagnetic storm caused the loss of 38 newly launched Starlink satellites due to increased atmospheric drag and radiation exposure (Weather.gov).

9.6 Risk Assessment and Preventive Measures

Risk Assessment

Understanding the risks posed by space weather to satellite operations is crucial for developing effective preventive measures. This involves analyzing historical data, modeling potential scenarios, and continuously monitoring space weather conditions.

Preventive Measures

- **Improved Forecasting:** Enhancing space weather forecasting capabilities to provide early warnings and allow for proactive measures.
- **Design Improvements:** Incorporating robust shielding, radiation-hardened components, and redundant systems in satellite design.
- **Operational Adjustments:** Implementing protocols for power management, orbit adjustments, and communication resilience during space weather events.

- **Collaboration and Data Sharing:** Strengthening international collaboration for data sharing, joint research, and coordinated responses to space weather threats.

9.7 Economic and Strategic Impacts

Economic Impacts

Space weather can have significant economic impacts on the satellite industry:

- **Satellite Replacement Costs:** Damage from space weather events can necessitate the replacement of expensive satellite hardware. For instance, the loss of 38 Starlink satellites in February 2022 due to increased atmospheric drag from a geomagnetic storm represents a substantial financial loss (Weather.gov).
- **Operational Costs:** Increased frequency of maneuvers to avoid collisions or adjust orbits during space weather events can drive up operational costs.
- **Service Disruptions:** Disruptions to communication, navigation, and other services provided by satellites can result in economic losses for businesses and consumers relying on these services.

Strategic Impacts

Beyond economic considerations, space weather can have strategic implications:

- **National Security:** Satellites play a critical role in national security operations, including surveillance, communication, and navigation. Disruptions caused by space weather can impact military readiness and strategic capabilities.
- **Global Positioning System (GPS):** GPS is essential for military operations, emergency response, and critical infrastructure. Space weather-induced errors in GPS positioning can have far-reaching consequences.

9.8 Future Research and Development

Advancements in Monitoring and Forecasting

Continued investment in space weather monitoring and forecasting is crucial for improving the resilience of satellite operations:

- **Next-Generation Satellites**: Developing new satellites equipped with advanced sensors to monitor space weather in real-time and provide early warnings.
- **Enhanced Models**: Improving space weather models to better predict the impacts of solar flares, CMEs, and geomagnetic storms on satellite operations.

Innovations in Satellite Design

Future satellite designs will need to incorporate advanced features to withstand the harsh space weather environment:

- **Radiation-Hardened Components**: Utilizing materials and technologies that are more resistant to radiation damage.
- **Adaptive Systems**: Designing satellites with systems that can adapt to changing space weather conditions, such as dynamic shielding and real-time fault correction.

International Collaboration and Policy Development

Strengthening international collaboration and developing comprehensive policies will be essential for addressing the global challenges posed by space weather:

- **Global Data Sharing**: Establishing frameworks for the sharing of space weather data among nations and organizations to improve global preparedness.
- **Space Traffic Management**: Developing international policies for space traffic management to ensure the safe and efficient operation of satellites during space weather events.

Public Awareness and Education

Raising awareness about space weather and its impacts on satellite operations is essential for fostering a resilient society:

- **Educational Programs**: Implementing educational programs to teach students and the general public about space weather and its effects on technology.
- **Public Outreach Campaigns**: Using media and public events to inform communities and industry stakeholders about space weather impacts and preparedness measures.

Satellites are critical for various applications, from communication and navigation to weather forecasting and scientific research. However, they are vulnerable to space weather events that can cause significant disruptions and damage. Understanding the impacts of space weather on satellite operations is essential for developing effective mitigation strategies and ensuring the reliability and resilience of these vital systems.

This chapter has provided a comprehensive overview of the impacts of space weather on satellite operations, including the mechanisms of radiation exposure, electrostatic charging, and geomagnetic storms. It has also discussed the specific vulnerabilities of different types of satellites and the strategies for monitoring, mitigating, and managing these impacts.

As we look to the future, continued research, international collaboration, and public awareness will be essential for addressing the challenges posed by space weather and ensuring the continued success of satellite operations. By investing in education, innovation, and global cooperation, we can better prepare for and mitigate the impacts of space weather on our increasingly satellite-dependent world.

1. **Understanding Vulnerabilities:** Recognizing the various ways space weather can impact satellite operations is crucial for developing effective mitigation strategies.
2. **Technological and Operational Solutions:** Implementing resilient technologies and adjusting operational procedures can help protect satellites from space weather impacts.
3. **Monitoring and Forecasting:** Continuous monitoring and accurate forecasting of space weather are essential for proactive risk management and mitigation.
4. **International Collaboration:** Global cooperation in space weather research, data sharing, and mitigation efforts is vital for safeguarding satellite operations.
5. **Public Awareness:** Educating the public and industry stakeholders about space weather impacts and preparedness measures is crucial for fostering a resilient society.

This comprehensive chapter on the impact of space weather on satellite operations sets the stage for understanding the subsequent chapters, which will explore specific space weather phenomena, their impacts on technology, and strategies for mitigation and preparedness.

CHAPTER 10: SPACE WEATHER AND HUMAN HEALTH

10.1 Introduction to Space Weather

Definition and Major Events

Space weather refers to the dynamic conditions in Earth's outer space environment influenced by solar activities such as solar flares, coronal mass ejections (CMEs), and geomagnetic storms. These events can release vast amounts of energy and particles, which can interact with Earth's magnetosphere and atmosphere, leading to various space weather phenomena.

Solar Flares and CMEs

Solar flares are intense bursts of radiation emanating from the Sun's surface, often associated with sunspots. CMEs involve the ejection of massive amounts of solar plasma and magnetic fields into space. When directed towards Earth, these events can significantly disturb the magnetosphere, leading to geomagnetic storms.

Geomagnetic Storms

Geomagnetic storms are major disturbances of Earth's magnetosphere that occur when there is a significant transfer of energy from the solar wind into the space environment surrounding Earth. These storms can cause widespread effects, including enhanced radiation levels and disruptions to technological systems.

10.2 Radiation Exposure and Human Health

Radiation Types and Sources

Space weather-related radiation primarily includes ionizing radiation from solar particle events (SPEs) and galactic cosmic rays (GCRs). SPEs are associated with solar flares and CMEs, while GCRs originate from outside the solar system and constantly bombard Earth and space environments.

Health Effects of Ionizing Radiation

Ionizing radiation poses significant health risks due to its ability to strip electrons from atoms, leading to cellular and DNA damage. The health effects include:

- **Acute Radiation Syndrome (ARS):** Characterized by nausea, vomiting, fatigue, and potential death following high doses of radiation over a short period.
- **Long-Term Health Risks:** Chronic exposure can lead to increased risks of cancer, cataracts, cardiovascular diseases, and neurological disorders.
- **DNA Damage:** Radiation can cause direct damage to DNA, resulting in mutations and increased cancer risk.

Deterministic and Stochastic Effects

Radiation exposure can result in deterministic and stochastic effects:

- **Deterministic Effects:** Occur after high doses of radiation and include ARS and cataract development. These effects have a threshold dose and can lead to severe symptoms if exceeded.
- **Stochastic Effects:** Long-term effects without a threshold dose, increasing in probability with the dose. The primary concern is cancer, with risks rising cumulatively over time.

10.3 Space Weather Impacts on Aviation and Astronaut Health

Radiation Risks in Aviation

Aviation personnel, especially those on high-altitude and polar routes, are at heightened risk of radiation exposure due to space weather:

- **Increased Radiation Dose:** SPEs can significantly elevate radiation doses received by flight crews and passengers, particularly on high-latitude flights.
- **Pregnant Crew Members:** Higher radiation exposure poses additional risks to pregnant crew members, potentially affecting fetal development.

Radiation Risks for Astronauts

Astronauts in low Earth orbit (LEO) and beyond face increased radiation exposure due to reduced protection from Earth's atmosphere and magnetosphere:

- **LEO Missions:** Astronauts on the International Space Station (ISS) experience higher radiation levels, especially during SPEs.
- **Deep Space Missions:** Missions beyond LEO, such as those to the Moon and Mars, involve greater radiation risks from prolonged exposure to GCRs and solar radiation.

Protective Measures for Aviation and Astronauts

Protective strategies include:

- **Radiation Monitoring:** Continuous monitoring using dosimeters to assess exposure levels and take timely protective actions.
- **Shielding Technologies:** Utilizing advanced materials for shielding, such as polyethylene and hydrogen-rich compounds, to protect against radiation.
- **Mission Planning:** Adjusting flight paths and mission schedules to avoid periods of high solar activity and reduce time spent in high-radiation zones.

10.4 Space Weather Impacts on Ground-Based Health Systems

Disruptions to Medical Technologies

Geomagnetic storms can disrupt medical technologies essential for healthcare:

- **Medical Imaging:** MRI machines and other imaging technologies can experience operational interference during geomagnetic storms, affecting diagnostic capabilities.
- **Electronic Medical Records:** Power grid and communication network disruptions can impact the reliability and accessibility of electronic medical records.

Impact on Public Health Infrastructure

Space weather events can indirectly affect public health infrastructure, leading to significant health implications:

- **Power Outages:** Geomagnetic storms can cause power outages, disrupting hospital operations and critical health services.
- **Communication Disruptions:** Interference with communication networks can hinder emergency response and coordination during health crises.

Preventive Measures for Ground-Based Systems

Ensuring resilience in health systems includes:

- **Backup Power Systems:** Hospitals and medical facilities need robust backup power systems to maintain operations during power outages.
- **Resilient Infrastructure:** Investing in infrastructure improvements to enhance the resilience of power grids and communication networks against geomagnetic disturbances.

10.5 Psychological and Behavioral Health Impacts

Mental Health Effects of Space Weather

Space weather can impact mental health, particularly for those in space environments:

- **Isolation and Confinement:** Astronauts face isolation and confinement during long missions, which can be exacerbated by communication disruptions due to space weather.
- **Sleep Disturbances:** Changes in geomagnetic activity can affect sleep patterns and circadian rhythms, leading to sleep disturbances and associated mental health issues.

Behavioral Health Strategies in Space Missions

Maintaining astronaut mental health is crucial for mission success:

- **Psychological Support:** Providing continuous psychological support and counseling services for astronauts.
- **Environmental Control:** Ensuring a stable and comfortable environment within spacecraft to mitigate the psychological impacts of space weather.

9.6 Monitoring and Mitigation Strategies

Monitoring Space Weather

Continuous monitoring is vital for protecting human health from space weather impacts:

- **Space-Based Observatories:** Satellites such as SOHO and ACE provide real-time data on solar activity and space weather conditions.
- **Ground-Based Observatories:** Magnetometers and ionosondes monitor geomagnetic activity and ionospheric conditions, offering comprehensive data for space weather analysis.

Radiation Mitigation Technologies

Developing advanced technologies to mitigate radiation exposure includes:

- **Advanced Shielding:** Researching new materials and technologies for effective radiation shielding, including polyethylene and hydrogen-rich compounds.
- **Pharmaceutical Interventions:** Exploring radioprotective drugs to mitigate radiation damage for astronauts and high-risk aviation personnel.

International Collaboration

Global cooperation is essential for effective space weather monitoring and mitigation:

- **Data Sharing:** Promoting the exchange of space weather data and forecasts among countries and organizations.
- **Joint Research Initiatives:** Collaborating on research projects to improve understanding and mitigation of space weather impacts on human health.

Public Awareness and Education

Raising public awareness about space weather impacts on human health is crucial:

- **Educational Programs:** Implementing educational programs to teach the public about space weather and its health effects.
- **Public Outreach:** Conducting outreach campaigns to inform communities and stakeholders about space weather impacts and preparedness measures.

Space weather significantly impacts human health through increased radiation exposure, disruptions to critical health technologies, and psychological stress. Understanding these impacts is essential for developing effective mitigation strategies and ensuring the well-being of individuals both on Earth and in space.

This chapter has provided a comprehensive overview of the ways space weather affects human health, including radiation exposure, impacts on aviation and astronaut health, disruptions to ground-based health systems, and psychological effects. It has also discussed strategies for monitoring, mitigating, and managing these impacts.

As we look to the future, continued research, international collaboration, and public awareness will be essential for addressing the challenges posed by space weather and ensuring the health and safety of individuals affected by these phenomena. By investing in education, innovation, and global cooperation, we can better prepare for and mitigate the impacts of space weather on human health.

1. **Radiation Exposure Risks:** Recognizing the health risks associated with radiation exposure from space weather events is crucial for developing protective measures.
2. **Impacts on Aviation and Astronaut Health:** Understanding how space weather affects aviation and astronaut health helps in planning and implementing effective protective strategies.
3. **Disruptions to Ground-Based Systems:** Geomagnetic storms can disrupt medical technologies and public health infrastructure, highlighting the need for resilient systems.
4. **Psychological Health:** Addressing the psychological impacts of space weather, especially for astronauts, is essential for maintaining overall health and mission success.
5. **Monitoring and Mitigation:** Continuous monitoring and advanced mitigation technologies are critical for protecting human health from space weather impacts.
6. **International Collaboration:** Global cooperation in space weather research, data sharing, and mitigation efforts is vital for safeguarding human health.

This comprehensive chapter on space weather and human health sets the stage for understanding the subsequent chapters, which will explore specific space weather phenomena, their impacts on technology, and strategies for mitigation and preparedness.

CHAPTER 11: EFFECTS ON COMMUNICATION SYSTEMS

11.1 Overview of Communication Systems

Types of Communication Systems

Communication systems affected by space weather include:

- **Satellite Communications:** Utilized for global telecommunications, television broadcasting, and internet services, satellites operate in geostationary and low Earth orbits.
- **High-Frequency (HF) Radio Communications:** Essential for long-distance communication in aviation, maritime, and military operations.
- **Global Navigation Satellite Systems (GNSS):** Systems like GPS, GLONASS, and Galileo provide critical positioning, navigation, and timing services.
- **Cellular and Wireless Networks:** Depend on a combination of ground-based infrastructure and satellite links to ensure global communication coverage.

Importance of Communication Systems

These systems are vital for:

- **Emergency Services:** Reliable communication during natural disasters and emergencies.
- **Military Operations:** Secure and dependable communication for defense and strategic operations.
- **Commercial Services:** Day-to-day business operations, financial transactions, and internet services.
- **Scientific Research:** Data transmission from remote sensing satellites and other scientific instruments.

11.2 Mechanisms of Space Weather Impacts on Communication Systems

Solar Flares

Solar flares emit bursts of radiation, including X-rays and extreme ultraviolet (EUV) radiation, affecting the ionosphere:

- **Ionospheric Disturbances:** Solar flares increase ionization in the D and E layers of the ionosphere, causing signal absorption and blackouts of HF radio waves.
- **Sudden Ionospheric Disturbances (SIDs):** Rapid changes in ionospheric conditions can lead to signal fading and loss of communication for HF radio systems.

Coronal Mass Ejections (CMEs)

CMEs eject solar plasma and magnetic fields into space, which can interact with the Earth's magnetosphere:

- **Geomagnetic Storms:** CMEs cause geomagnetic storms, leading to enhanced ionospheric currents and fluctuations that disrupt HF radio communication and GNSS signals.
- **Plasma Density Irregularities:** CMEs introduce irregularities in ionospheric plasma density, affecting satellite communication and GNSS accuracy.

Geomagnetic Storms

Geomagnetic storms, triggered by CMEs and high-speed solar wind streams, significantly impact communication systems:

- **Ionospheric Scintillation:** Rapid changes in ionospheric electron density cause signal scattering and fading, particularly affecting GNSS signals and satellite communications.
- **Geomagnetically Induced Currents (GICs):** These currents can induce voltage fluctuations and outages in communication infrastructure, affecting both terrestrial and satellite systems.

11.3 Impacts on Specific Communication Systems

Satellite Communications

Satellite communications are highly susceptible to space weather:

- **Signal Degradation and Loss:** Increased ionization and plasma density irregularities in the ionosphere can cause signal delays, bending, and fading.

- **Satellite Damage:** High-energy particles from solar flares and CMEs can damage satellite electronics and solar panels, leading to operational anomalies and potential failures.
- **Orbital Changes:** Increased atmospheric drag during geomagnetic storms can alter satellite orbits, affecting their positioning and communication capabilities.

High-Frequency (HF) Radio Communications

HF radio communications are particularly vulnerable to ionospheric disturbances:

- **Absorption and Blackouts:** Increased ionization in the D-layer during solar flares can absorb HF signals, causing communication blackouts.
- **Frequency Shifts:** Changes in ionospheric conditions can cause frequency shifts, affecting the clarity and reliability of HF radio transmissions.

Global Navigation Satellite Systems (GNSS)

GNSS systems, including GPS, are critical for positioning, navigation, and timing:

- **Signal Delays and Errors:** Ionospheric scintillation and density irregularities cause signal delays and positional errors, reducing GNSS accuracy.
- **Loss of Lock:** Severe ionospheric disturbances can cause GNSS receivers to lose lock on satellite signals, disrupting navigation and timing services.

Cellular and Wireless Networks

While less directly impacted, cellular and wireless networks can still be affected by space weather:

- **Satellite Link Disruptions:** Cellular networks that rely on satellite links for backhaul services can experience disruptions during space weather events.
- **Infrastructure Damage:** Geomagnetically induced currents can damage ground-based communication infrastructure, affecting service reliability.

11.4 Historical Case Studies of Space Weather Impacts

The Carrington Event (1859)

The Carrington Event was the most powerful geomagnetic storm on record, causing widespread disruptions:

- **Telegraph Systems:** Induced currents caused telegraph systems to fail and operators to receive electric shocks.
- **Auroras:** Auroras were visible at low latitudes, indicating the intensity of the storm.

The March 1989 Geomagnetic Storm

This storm caused significant disruptions in communication systems:

- **Quebec Power Grid Collapse:** Geomagnetically induced currents caused a major power outage, affecting communication services.
- **Satellite Anomalies:** Several satellites experienced operational anomalies due to increased radiation levels and ionospheric disturbances.

The Halloween Storms (2003)

A series of intense solar storms in October-November 2003 led to:

- **GNSS Disruptions:** Significant errors in GPS positioning, affecting aviation and maritime navigation.
- **Satellite Failures:** Communication satellites experienced increased anomalies and temporary outages.

11.5 Monitoring and Forecasting Space Weather

Space-Based Observatories

Space-based observatories provide critical data for monitoring and forecasting space weather:

- **SOHO (Solar and Heliospheric Observatory):** Monitors solar activity, providing early warnings of solar flares and CMEs.
- **ACE (Advanced Composition Explorer):** Measures solar wind and cosmic rays near the Lagrange point L1, providing real-time data on space weather conditions.

- **Parker Solar Probe**: Studies the solar corona and solar wind to improve understanding and prediction of space weather events.

Ground-Based Observatories

Ground-based observatories complement space-based observations:

- **Magnetometers**: Measure variations in Earth's magnetic field, providing data on geomagnetic storms and substorms.
- **Ionosondes**: Measure ionospheric properties, helping to assess the impact of space weather on communications.

Forecasting Techniques

Advancements in forecasting techniques enhance our ability to predict space weather impacts:

- **Empirical Models**: Use historical data to predict the likelihood and severity of space weather events.
- **Machine Learning and AI**: Analyze large datasets to identify patterns and improve prediction accuracy.

11.6 Mitigation Strategies for Communication Systems

Technological Solutions

Developing resilient technologies is crucial for mitigating space weather impacts:

- **Radiation-Hardened Electronics**: Incorporating radiation-hardened components to withstand high-energy particle radiation.
- **Adaptive Systems**: Designing communication systems that can adapt to changing ionospheric conditions, such as frequency agility in HF radio systems.

Adjusting operational procedures can help minimize the risks posed by space weather:

- **Frequency Management**: Using multiple frequencies and adjusting them based on ionospheric conditions to maintain HF radio communication.

- **Satellite Maneuvers:** Adjusting satellite orbits and operations to minimize exposure to radiation belts and high-radiation areas during space weather events.
- **Power Management:** Reducing power loads and temporarily shutting down non-essential systems during high-radiation periods to protect sensitive components in satellites.

Infrastructure Improvements

Improving the resilience of communication infrastructure is essential for mitigating space weather impacts:

- **Enhanced Ground Infrastructure:** Strengthening ground-based communication infrastructure to withstand geomagnetic storms and their induced currents.
- **Redundant Systems:** Developing redundant communication pathways to ensure continuity of services during space weather events.
- **Hardening Communication Networks:** Implementing protective measures to shield critical infrastructure from geomagnetically induced currents (GICs).

Policy and Regulation

Governments and regulatory bodies play a crucial role in mitigating the effects of space weather on communication systems:

- **International Standards:** Developing and enforcing international standards for space weather resilience in communication systems.
- **Regulatory Frameworks:** Establishing regulatory frameworks that mandate the implementation of space weather mitigation strategies by communication service providers.
- **Emergency Preparedness Plans:** Requiring communication service providers to develop and maintain comprehensive emergency preparedness plans that include space weather contingencies.

11.7 Economic and Strategic Impacts of Space Weather on Communication Systems

Economic Impacts

Space weather can have significant economic implications for communication systems:

- **Service Disruptions:** Interruptions in communication services can lead to substantial economic losses for businesses and consumers. For example, disruptions to GNSS can affect transportation and logistics, leading to delays and increased operational costs.
- **Infrastructure Damage:** Geomagnetically induced currents can damage communication infrastructure, necessitating costly repairs and replacements.
- **Insurance Costs:** Increased risk of space weather-related damages can lead to higher insurance premiums for communication service providers.

Strategic Impacts

Beyond economic considerations, space weather impacts on communication systems can have strategic implications:

- **National Security:** Reliable communication systems are critical for national security operations, including military communication and coordination. Space weather-induced disruptions can compromise these operations.
- **Global Navigation and Timing:** GNSS systems are essential for global navigation and timing, which are critical for various sectors, including finance, transportation, and emergency services. Disruptions to GNSS can have far-reaching strategic consequences.

10.8 Future Directions in Space Weather Research and Mitigation

Advancements in Monitoring and Forecasting

Continued advancements in space weather monitoring and forecasting are essential for improving resilience:

- **Next-Generation Satellites:** Developing new satellites equipped with advanced sensors to monitor space weather in real-time and provide early warnings.
- **Enhanced Forecast Models:** Improving space weather forecast models to increase prediction accuracy and provide longer lead times for mitigation measures.

Innovations in Technology

Technological innovations can enhance the resilience of communication systems to space weather:

- **Smart Antennas:** Developing smart antenna systems that can dynamically adjust to changing ionospheric conditions and maintain communication links.
- **Quantum Communication:** Exploring the potential of quantum communication technologies, which may offer greater resilience to space weather impacts.

International Collaboration and Policy Development

Strengthening international collaboration and developing comprehensive policies will be essential for addressing the global challenges posed by space weather:

- **Global Data Sharing Networks:** Establishing global networks for sharing space weather data and forecasts to improve collective preparedness.
- **International Policy Frameworks:** Developing international policy frameworks that promote the adoption of best practices and standards for space weather resilience in communication systems.

Public Engagement and Education

Engaging the public and stakeholders is crucial for building a resilient society:

- **Awareness Campaigns:** Conducting awareness campaigns to inform the public and industry stakeholders about space weather risks and mitigation strategies.
- **Educational Programs:** Integrating space weather education into school curricula and professional training programs to build a knowledgeable and prepared workforce.

Space weather significantly impacts communication systems, from satellite communications to HF radio and GNSS. Understanding these impacts is essential for developing effective mitigation strategies and ensuring the reliability and resilience of these critical systems.

This chapter has provided a comprehensive overview of the ways space weather affects communication systems, including the mechanisms of disruption, historical case studies, and strategies for monitoring, mitigating, and managing these impacts. As we look to the future, continued research, international collaboration, and public awareness will be essential for addressing the challenges posed by space weather and ensuring the continued success of communication systems.

1. **Understanding Vulnerabilities:** Recognizing the various ways space weather can impact communication systems is crucial for developing effective mitigation strategies.
2. **Technological and Operational Solutions:** Implementing resilient technologies and adjusting operational procedures can help protect communication systems from space weather impacts.
3. **Monitoring and Forecasting:** Continuous monitoring and accurate forecasting of space weather are essential for proactive risk management and mitigation.
4. **International Collaboration:** Global cooperation in space weather research, data sharing, and mitigation efforts is vital for safeguarding communication systems.
5. **Public Awareness:** Educating the public and industry stakeholders about space weather impacts and preparedness measures is crucial for fostering a resilient society.

This comprehensive chapter on space weather and communication systems sets the stage for understanding the subsequent chapters, which will explore specific space weather phenomena, their impacts on technology, and strategies for mitigation and preparedness.

CHAPTER 12: SPACE WEATHER AND POWER GRIDS

Space weather, including solar flares, coronal mass ejections (CMEs), and geomagnetic storms, poses significant risks to national infrastructure, particularly power grids. This chapter explores how geomagnetic storms can disrupt power grids, the underlying mechanisms, historical impacts, and strategies for monitoring, mitigating, and managing these effects.

12.1 Overview of Space Weather and Power Grids

Introduction to Space Weather

Space weather refers to the environmental conditions in space influenced by solar activities such as solar flares, CMEs, and geomagnetic storms. These events can release vast amounts of energy and particles, which interact with Earth's magnetosphere and atmosphere, leading to various space weather phenomena.

Importance of Power Grids

Power grids are critical infrastructure for modern society, providing electricity for homes, businesses, healthcare, transportation, and communication systems. The stability and reliability of power grids are essential for national security, economic stability, and public safety.

12.2 Mechanisms of Space Weather Impacts on Power Grids

Geomagnetic Induced Currents (GICs)

Geomagnetic storms can induce electric currents in the Earth's surface, known as geomagnetic induced currents (GICs). These currents can enter power grids through grounded transformers and flow through power lines and infrastructure:

- **Transformer Saturation:** GICs can cause transformer cores to saturate, leading to overheating, increased reactive power consumption, and potential damage to the transformer.
- **Voltage Instability:** GICs can cause voltage instability and fluctuations, impacting the overall stability of the power grid.
- **Harmonics and Resonance:** GICs can introduce harmonics into the power system, causing equipment malfunctions and increased losses.

Transformer Damage

Transformers are particularly vulnerable to GICs due to their design and operation:

- **Thermal Stress:** GICs can cause localized heating and thermal stress within transformer windings, leading to insulation breakdown and potential failure.
- **Magnetostriction:** GICs can cause mechanical stress in transformer cores due to magnetostriction, resulting in vibrations and noise that can damage internal components.

System-Wide Disruptions

Geomagnetic storms can cause widespread disruptions in power grids:

- **Blackouts:** Severe geomagnetic storms can cause large-scale blackouts by destabilizing the power grid and causing cascading failures.
- **Grid Instability:** GICs can cause grid instability by affecting voltage regulation and reactive power balance, leading to potential grid collapse.

12.3 Historical Case Studies of Geomagnetic Storm Impacts on Power Grids

The March 1989 Geomagnetic Storm

The March 1989 geomagnetic storm is one of the most significant events in recent history, causing extensive damage to power grids:

- **Quebec Blackout:** GICs induced by the storm caused a major power outage in Quebec, Canada, leaving millions of people without power for several hours.
- **Transformer Damage:** The storm caused damage to transformers in Canada and the United States, resulting in costly repairs and replacements.

The Halloween Storms (2003)

The Halloween storms of October-November 2003 also had a significant impact on power grids:

- **Transformer Failures:** GICs caused by the storms led to transformer failures and power disruptions in Sweden and South Africa.
- **Increased Awareness:** The events highlighted the vulnerability of power grids to geomagnetic storms and prompted increased research and mitigation efforts.

The Carrington Event (1859)

The Carrington Event, the most powerful geomagnetic storm on record, provides insights into the potential impact of extreme space weather events:

- **Telegraph System Failures:** Induced currents caused telegraph systems to fail and operators to receive electric shocks.
- **Auroras:** Auroras were visible at low latitudes, indicating the intensity of the storm. Although modern power grids did not exist at the time, similar events today could cause catastrophic disruptions.

12.4 Monitoring and Forecasting Space Weather

Space-Based Observatories

Space-based observatories provide critical data for monitoring and forecasting space weather:

- **SOHO (Solar and Heliospheric Observatory):** Monitors solar activity, providing early warnings of solar flares and CMEs.
- **ACE (Advanced Composition Explorer):** Measures solar wind and cosmic rays near the Lagrange point L1, providing real-time data on space weather conditions.
- **Parker Solar Probe:** Studies the solar corona and solar wind to improve understanding and prediction of space weather events.

Ground-Based Observatories

Ground-based observatories complement space-based observations:

- **Magnetometers:** Measure variations in Earth's magnetic field, providing data on geomagnetic storms and substorms.
- **Ionosondes:** Measure ionospheric properties, helping to assess the impact of space weather on power grids.

Forecasting Techniques

Advancements in forecasting techniques enhance our ability to predict space weather impacts:

- **Empirical Models:** Use historical data to predict the likelihood and severity of space weather events.
- **Machine Learning and AI:** Analyze large datasets to identify patterns and improve prediction accuracy.

12.5 Mitigation Strategies for Power Grids

Technological Solutions

Developing resilient technologies is crucial for mitigating space weather impacts on power grids:

- **GIC Blocking Devices:** Installing devices to block or reduce GICs entering the power grid, protecting transformers and other infrastructure.
- **Advanced Transformers:** Designing transformers with increased tolerance to GICs, such as those with higher thermal and mechanical stress resistance.
- **Monitoring Systems:** Implementing real-time monitoring systems to detect GICs and other anomalies, allowing for rapid response and mitigation.

Operational Strategies

Adjusting operational procedures can help minimize the risks posed by space weather:

- **Load Management:** Adjusting load distribution and managing power flows to reduce the impact of GICs on critical infrastructure.
- **Transformer Switching:** Temporarily switching off or reconfiguring transformers during severe geomagnetic storms to prevent damage.
- **Reactive Power Compensation:** Using reactive power compensation techniques to maintain voltage stability and prevent grid collapse.

Infrastructure Improvements

Improving the resilience of power grid infrastructure is essential for mitigating space weather impacts:

- **Enhanced Grounding:** Implementing enhanced grounding techniques to safely dissipate GICs and reduce their impact on the power grid.
- **Redundant Systems:** Developing redundant power pathways and infrastructure to ensure continuity of service during space weather events.
- **Hardening Communication Networks:** Implementing protective measures to shield critical infrastructure from geomagnetically induced currents (GICs).

Policy and Regulation

Governments and regulatory bodies play a crucial role in mitigating the effects of space weather on power grids:

- **International Standards:** Developing and enforcing international standards for space weather resilience in power grids.
- **Regulatory Frameworks:** Establishing regulatory frameworks that mandate the implementation of space weather mitigation strategies by power grid operators.
- **Emergency Preparedness Plans:** Requiring power grid operators to develop and maintain comprehensive emergency preparedness plans that include space weather contingencies.

12.6 Economic and Strategic Impacts of Space Weather on Power Grids

Economic Impacts

Space weather can have significant economic implications for power grids:

- **Service Disruptions:** Interruptions in power supply can lead to substantial economic losses for businesses and consumers. For example, disruptions to industrial operations can result in production delays and financial losses.
- **Infrastructure Damage:** GIC-induced damage to transformers and other infrastructure can result in costly repairs and replacements.
- **Insurance Costs:** Increased risk of space weather-related damages can lead to higher insurance premiums for power grid operators.

Strategic Impacts

Beyond economic considerations, space weather impacts on power grids can have strategic implications:

- **National Security:** Reliable power grids are critical for national security operations, including military communication and coordination. Space weather-induced disruptions can compromise these operations.
- **Public Safety:** Power outages can impact public safety by disrupting emergency services, healthcare, and critical infrastructure.
- **Economic Stability:** Prolonged power outages can affect economic stability by disrupting financial transactions, supply chains, and essential services.

12.7 Future Directions in Space Weather Research and Mitigation

Advancements in Monitoring and Forecasting

Continued advancements in space weather monitoring and forecasting are essential for improving resilience:

- **Next-Generation Satellites:** Developing new satellites equipped with advanced sensors to monitor space weather in real-time and provide early warnings.
- **Enhanced Forecast Models:** Improving space weather forecast models to increase prediction accuracy and provide longer lead times for mitigation measures.

Innovations in Technology

Technological innovations can enhance the resilience of power grids to space weather:

- **Smart Grid Technologies:** Developing smart grid technologies that can dynamically respond to space weather-induced disruptions and maintain stability.
- **Energy Storage Systems:** Implementing advanced energy storage systems to provide backup power during space weather-induced outages.
- **Microgrids:** Developing microgrids that can operate independently of the main power grid during disruptions, ensuring continuity of critical services.

International Collaboration and Policy Development

Strengthening international collaboration and developing comprehensive policies will be essential for addressing the global challenges posed by space weather:

- **Global Data Sharing Networks:** Establishing global networks for sharing space weather data and forecasts to improve collective preparedness and response.
- **International Policy Frameworks:** Developing international policy frameworks that promote the adoption of best practices and standards for space weather resilience in power grids. This includes collaboration through organizations like the International Council on Large Electric Systems (CIGRÉ) and the International Electrotechnical Commission (IEC).

Public Engagement and Education

Engaging the public and stakeholders is crucial for building a resilient society:

- **Awareness Campaigns:** Conducting awareness campaigns to inform the public and industry stakeholders about space weather risks and mitigation strategies. Utilizing media platforms, workshops, and seminars can effectively disseminate information.
- **Educational Programs:** Integrating space weather education into school curricula and professional training programs to build a knowledgeable and prepared workforce. Universities and institutions can include space weather courses in their curricula to prepare future engineers and scientists (USGS) (MDPI).

Space weather significantly impacts power grids through geomagnetic induced currents (GICs) and associated disruptions. Understanding these impacts is essential for developing effective mitigation strategies and ensuring the reliability and resilience of power grids.

This chapter has provided a comprehensive overview of the ways space weather affects power grids, including the mechanisms of disruption, historical case studies, and strategies for monitoring, mitigating, and managing these impacts. As we look to the future, continued research, international collaboration, and public awareness will be essential for addressing the challenges posed by space weather and ensuring the continued success of power grids.

1. **Understanding Vulnerabilities:** Recognizing the various ways space weather can impact power grids is crucial for developing effective mitigation strategies.
2. **Technological and Operational Solutions:** Implementing resilient technologies and adjusting operational procedures can help protect power grids from space weather impacts.
3. **Monitoring and Forecasting:** Continuous monitoring and accurate forecasting of space weather are essential for proactive risk management and mitigation.
4. **International Collaboration:** Global cooperation in space weather research, data sharing, and mitigation efforts is vital for safeguarding power grids.
5. **Public Awareness:** Educating the public and industry stakeholders about space weather impacts and preparedness measures is crucial for fostering a resilient society.

This comprehensive chapter on space weather and power grids sets the stage for understanding the subsequent chapters, which will explore specific space weather phenomena, their impacts on technology, and strategies for mitigation and preparedness.

CHAPTER 13: MITIGATION STRATEGIES FOR SPACE WEATHER

Space weather poses significant risks to various technological systems and human activities, including power grids, communication systems, aviation, and space operations. Effective mitigation strategies are essential to minimize these risks and enhance resilience against space weather events. This chapter explores detailed mitigation strategies, covering technological solutions, operational procedures, policy frameworks, and international collaboration.

13.1 Overview of Space Weather Mitigation

Importance of Mitigation

Mitigating space weather impacts is crucial for protecting critical infrastructure, ensuring public safety, and maintaining economic stability. The increasing reliance on technology makes societies more vulnerable to space weather effects, necessitating comprehensive mitigation strategies.

Key Areas of Focus

- **Technological Solutions:** Developing advanced technologies to protect infrastructure.
- **Operational Strategies:** Implementing procedures to manage space weather risks.
- **Policy and Regulation:** Establishing frameworks to enforce mitigation measures.
- **International Collaboration:** Promoting global cooperation in space weather monitoring and response.

13.2 Technological Solutions

Radiation-Hardened Electronics

Developing radiation-hardened electronics is essential for protecting space-based and ground-based systems:

- **Satellite Protection:** Using radiation-hardened components in satellites to withstand high-energy particles from solar flares and cosmic rays.
- **Power Grid Equipment:** Incorporating hardened components in power grid infrastructure to prevent damage from geomagnetic induced currents (GICs).

Advanced Shielding Materials

Utilizing advanced materials for shielding can protect critical systems from space weather impacts:

- **Satellite Shielding**: Implementing materials such as polyethylene and hydrogen-rich compounds to shield satellites from radiation.
- **Aircraft Shielding**: Enhancing shielding in aircraft to protect passengers and crew from increased radiation during solar particle events (SPEs).

Real-Time Monitoring Systems

Real-time monitoring systems are vital for detecting space weather events and responding promptly:

- **Space-Based Observatories**: Satellites like SOHO and ACE provide real-time data on solar activity.
- **Ground-Based Observatories**: Networks of magnetometers and ionosondes monitor geomagnetic activity and ionospheric conditions.

Automated Response Systems

Automated response systems can quickly implement protective measures during space weather events:

- **Power Grid Management**: Automated systems can adjust load distribution and manage reactive power to prevent grid instability during geomagnetic storms.
- **Satellite Operations**: Automated systems can maneuver satellites to minimize exposure to high-radiation areas.

13.3 Operational Strategies

Load Management in Power Grids

Effective load management can mitigate the impact of geomagnetic storms on power grids:

- **Load Redistribution**: Adjusting load distribution to reduce stress on critical components during geomagnetic storms.
- **Transformer Switching**: Temporarily switching off or reconfiguring transformers to prevent damage from GICs.

Frequency Management in Communications

Managing frequencies in communication systems can minimize disruptions:

- **HF Radio Communications:** Using multiple frequencies and adjusting them based on ionospheric conditions to maintain communication links.
- **GNSS Signal Integrity:** Implementing techniques to enhance GNSS signal integrity and reduce errors caused by ionospheric disturbances.

Mission Planning for Space Operations

Planning space missions to avoid periods of high solar activity can reduce risks:

- **Launch Scheduling:** Scheduling launches during periods of low solar activity to minimize exposure to radiation.
- **Mission Trajectories:** Designing mission trajectories to avoid regions of high radiation, such as the Van Allen belts.

Aviation Procedures

Implementing aviation procedures to protect passengers and crew from radiation exposure:

- **Flight Path Adjustments:** Rerouting flights, especially polar routes, during solar particle events to minimize radiation exposure.
- **Altitude Management:** Adjusting flight altitudes to reduce exposure to increased radiation during solar flares.

13.4 Policy and Regulation

International Standards

Developing and enforcing international standards for space weather mitigation is crucial:

- **IEC and ISO Standards:** Establishing standards for space weather resilience in various sectors, including power grids and telecommunications.
- **Compliance and Certification:** Ensuring compliance with international standards through certification processes.

National Regulatory Frameworks

National regulatory frameworks can mandate the implementation of mitigation measures:

- **Government Policies:** Governments can develop policies that require critical infrastructure operators to implement space weather mitigation strategies.
- **Emergency Preparedness Plans:** Requiring operators to develop and maintain emergency preparedness plans that include space weather contingencies.

Funding and Incentives

Providing funding and incentives for research and implementation of mitigation strategies:

- **Research Grants:** Funding research on advanced technologies and strategies for space weather mitigation.
- **Incentive Programs:** Offering incentives for industries to adopt and implement mitigation measures.

13.5 International Collaboration

Global Data Sharing Networks

Establishing global networks for sharing space weather data and forecasts:

- **ISES and INTERMAGNET:** International organizations that facilitate data sharing and collaboration in space weather monitoring.
- **Multinational Research Projects:** Collaborating on research projects to improve understanding and prediction of space weather events.

Joint Response Initiatives

Coordinating joint response initiatives to enhance global resilience:

- **International Drills:** Conducting international drills to practice coordinated response to space weather events.
- **Mutual Aid Agreements:** Establishing agreements for mutual aid and support during severe space weather events.

Capacity Building

Building capacity in developing countries to enhance global resilience:

- **Training Programs:** Providing training and education on space weather monitoring and mitigation.
- **Technology Transfer:** Facilitating the transfer of advanced technologies and best practices to developing countries.

13.6 Case Studies in Mitigation

Power Grid Resilience in North America

The North American power grid has implemented various mitigation strategies to enhance resilience:

- **GIC Monitoring Systems:** Installing GIC monitoring systems to detect and respond to geomagnetic storms.
- **Operational Protocols:** Developing operational protocols to manage load and protect transformers during geomagnetic storms.

Aviation Procedures in Polar Regions

Aviation authorities have implemented procedures to protect flights in polar regions:

- **Flight Path Adjustments:** Rerouting polar flights during solar particle events to minimize radiation exposure.
- **Radiation Monitoring:** Equipping aircraft with radiation monitors to assess exposure levels and make informed decisions.

Satellite Shielding Technologies

The satellite industry has developed advanced shielding technologies to protect against radiation:

- **Material Innovations:** Using materials like polyethylene and hydrogen-rich compounds for satellite shielding.
- **Design Improvements:** Incorporating shielding into satellite design to protect sensitive electronics and systems.

Mitigating the impacts of space weather requires a comprehensive approach that includes technological solutions, operational strategies, policy frameworks, and international collaboration. By implementing these strategies, we can enhance the resilience of critical infrastructure and protect public safety and economic stability.

This chapter has provided a detailed overview of mitigation strategies for space weather, covering various sectors and approaches. As we continue to advance our understanding of space weather, ongoing research, innovation, and cooperation will be essential for developing effective and sustainable mitigation strategies.

1. **Technological Solutions**: Developing advanced technologies such as radiation-hardened electronics and real-time monitoring systems is crucial for mitigating space weather impacts.
2. **Operational Strategies**: Implementing effective operational procedures, such as load management and frequency adjustments, can minimize disruptions during space weather events.
3. **Policy and Regulation**: Establishing and enforcing international standards and national regulatory frameworks can ensure the implementation of mitigation measures.
4. **International Collaboration**: Promoting global cooperation in data sharing, research, and response initiatives enhances resilience against space weather.
5. **Public Awareness**: Educating the public and stakeholders about space weather risks and mitigation strategies is essential for fostering a resilient society.

This comprehensive chapter on mitigation strategies for space weather sets the stage for understanding subsequent chapters, which will explore specific space weather phenomena, their impacts on technology, and additional strategies for preparedness and response

Space weather forecasting is crucial for mitigating the impacts of solar activity on technology and human activities. This chapter explores the latest forecasting methods, the application of advanced technologies, and the purpose and results of early forecasting efforts, integrating information from multiple sources to provide a comprehensive overview.

14.1 Introduction to Space Weather Forecasting

Importance of Forecasting

Forecasting space weather is essential for:

- **Protecting Infrastructure:** Safeguarding power grids, communication systems, and satellite operations from disruptions.
- **Ensuring Safety:** Shielding aviation and space missions from harmful radiation.
- **Economic Stability:** Preventing economic losses due to technological failures.

Key Objectives

The primary objectives include:

- **Early Warning:** Providing timely alerts about upcoming space weather events.
- **Impact Assessment:** Evaluating potential impacts on systems and preparing mitigation strategies.
- **Continuous Monitoring:** Ensuring real-time monitoring of solar and geomagnetic activities for accurate forecasts.

14.2 Latest Forecasting Methods

Empirical Models

Empirical models rely on historical data to predict space weather events:

- **Climatology Models:** Use long-term statistical data for probabilistic forecasts.
- **Persistence Models:** Predict that current conditions will continue for a short period.

Numerical Models

Numerical models simulate the physical processes of the Sun-Earth system:

- **Magnetohydrodynamic (MHD) Models:** Simulate plasma behavior in the Sun's atmosphere and solar wind. Examples include the WSA-Enlil model and the Space Weather Modeling Framework (SWMF) (Space.com).
- **Solar Wind Models:** Predict the arrival of solar wind structures, such as CMEs, at Earth by simulating their propagation through the heliosphere.

Machine Learning and AI

Machine learning and AI techniques are increasingly used to improve the accuracy of space weather forecasts:

- **Deep Learning Models:** Analyze large datasets of solar and geomagnetic activity to identify patterns. These models can learn from historical events to improve forecast accuracy.
- **Neural Networks:** Used to predict the occurrence of solar flares and geomagnetic storms based on real-time data from solar observatories (Phys.org) (Space.com).

Hybrid Models

Hybrid models combine empirical, numerical, and AI-based approaches to leverage the strengths of each method:

- **Integrated Forecasting Systems:** Use a combination of different models to provide comprehensive space weather forecasts. These systems can integrate data from multiple sources and use various modeling techniques to enhance forecast accuracy (Space.com).

14.3 Application of Technology in Forecasting

Space-Based Observatories

Space-based observatories play a crucial role in monitoring and forecasting space weather:

- **GOES-R Series Satellites:** Provide high-resolution data on solar activity and space weather events, enhancing forecasting capabilities (Space.com).
- **SOHO (Solar and Heliospheric Observatory):** Monitors solar activity, providing early warnings of solar flares and CMEs.
- **Advanced Composition Explorer (ACE):** Measures solar wind parameters and interplanetary magnetic fields, providing data for predicting geomagnetic storms.
- **Parker Solar Probe:** Studies the Sun's corona and solar wind, enhancing our understanding of the solar processes that drive space weather (Space.com).

Ground-Based Observatories

Ground-based observatories complement space-based data:

- **Magnetometers:** Measure geomagnetic field variations to detect geomagnetic storms.
- **Ionosondes:** Measure ionospheric properties, aiding in HF radio communication forecasts.

Real-Time Data Networks

Real-time data networks ensure continuous monitoring:

- **Global Positioning System (GPS) Networks:** Monitor ionospheric disturbances that affect GNSS accuracy.
- **Global Magnetometer Network:** Provides real-time geomagnetic activity data worldwide.

Supercomputing and Data Analysis

Advanced computing is crucial for processing large datasets:

- **Supercomputers:** Run high-resolution numerical simulations.
- **Big Data Analytics:** Utilize data mining and machine learning for improved forecasts (Space.com).

14.4 Purpose and Results of Early Forecasting

Early Warning Systems

Early warning systems provide timely alerts:

- **Alerts and Warnings:** Issued for solar flares, CMEs, and geomagnetic storms to allow for preventive measures to be taken.
- **Communication Channels:** Information is disseminated through various channels, including websites, email alerts, and mobile apps.

Impact Assessment and Preparedness

Early forecasting helps assess impacts and prepare responses:

- **Risk Assessment:** Evaluates potential impacts on critical infrastructure.
- **Preparedness Measures:** Implements preventive actions like adjusting satellite orbits, rerouting flights, and managing power grid loads.

Mitigation and Response

Forecasting enables timely mitigation and response efforts:

- **Infrastructure Protection:** Uses protective measures such as GIC blocking devices and enhanced shielding for satellites.
- **Operational Adjustments:** Modifies procedures to reduce risks, such as changing flight paths and communication frequencies.

14.5 Case Studies and Examples

The March 1989 Geomagnetic Storm

- **Forecasting Success:** Early warnings helped mitigate impacts on power grids and communication systems.
- **Impact:** The storm caused significant damage, including a major blackout in Quebec, highlighting the need for improved forecasting and mitigation strategies.

The Halloween Storms (2003)

- **Forecasting Improvements:** Enhanced forecasting methods and real-time data networks provided timely alerts for the Halloween storms.

- **Impact and Response:** Preventive measures were implemented, such as adjusting satellite operations and managing power grid loads, reducing the overall impact of the storms.

Recent Advances

- **Solar Cycle 24/25 Transition:** Forecasting methods have improved significantly during the transition between Solar Cycles 24 and 25, providing better predictions and early warnings for solar activity.
- **Integration of AI:** The integration of AI and machine learning techniques has led to more accurate and reliable forecasts, helping to mitigate the impacts of space weather events (Phys.org) (Space.com) (Space.com).

14.6 Future Directions in Space Weather Forecasting

Advancements in Modeling

Future advancements in space weather forecasting will focus on improving modeling techniques:

- **Higher Resolution Models:** Developing higher resolution numerical models to provide more detailed and accurate forecasts.
- **Ensemble Forecasting:** Using ensemble forecasting techniques to account for uncertainties and provide probabilistic forecasts.

Enhanced Observational Capabilities

Improving observational capabilities will enhance our ability to monitor and forecast space weather:

- **Next-Generation Satellites:** Launching new satellites with advanced sensors to monitor solar activity and space weather in real time.
- **Global Observing Networks:** Expanding global observing networks to improve data coverage and enhance forecasting accuracy.

Integration of Emerging Technologies

Emerging technologies will play a crucial role in advancing space weather forecasting:

- **Quantum Computing:** Utilizing quantum computing to process large datasets and run complex models more efficiently.

- **Internet of Things (IoT):** Integrating IoT devices for real-time monitoring of space weather impacts on critical infrastructure.

International Collaboration

Strengthening international collaboration will be essential for improving space weather forecasting:

- **Data Sharing Agreements:** Establishing agreements for sharing space weather data and forecasts among countries and organizations.
- **Joint Research Initiatives:** Collaborating on research projects to develop advanced forecasting techniques and improve global resilience (Phys.org) (Space.com).

Space weather forecasting is crucial for mitigating the impacts of solar activity on technological systems and human activities. This chapter has provided a comprehensive overview of the latest forecasting methods, the application of advanced technology, and the purpose and results of early forecasting efforts.

1. **Empirical and Numerical Models:** Using historical data and physical simulations to predict space weather events.
2. **Machine Learning and AI:** Leveraging advanced algorithms to improve forecast accuracy.
3. **Technological Integration:** Applying space-based and ground-based observatories, real-time data networks, and supercomputing for continuous monitoring and forecasting.
4. **Early Warning Systems:** Providing timely alerts to enable preventive measures and minimize impacts.
5. **Future Directions:** Advancing modeling techniques, enhancing observational capabilities, integrating emerging technologies, and promoting international collaboration.

This comprehensive chapter on forecasting space weather sets the stage for understanding the subsequent chapters, which will explore specific space weather phenomena, their impacts on technology, and additional strategies for preparedness and response.

pace weather monitoring systems are critical for detecting and forecasting space weather events. These systems utilize a combination of space-based and ground-based instruments to provide comprehensive data on solar activity, geomagnetic disturbances, and their potential impacts on Earth. This chapter explores the various components, technologies, and methodologies involved in space weather monitoring systems, highlighting the importance of these systems in mitigating the effects of space weather.

15.1 Introduction to Space Weather Monitoring

Purpose of Monitoring Systems

Space weather monitoring systems are designed to:

- **Detect and Track Solar Activity:** Monitor solar flares, coronal mass ejections (CMEs), and other solar phenomena.
- **Assess Geomagnetic Activity:** Measure disturbances in Earth's magnetic field and ionosphere.
- **Provide Early Warnings:** Issue alerts for potential space weather events that could impact technology and human activities.
- **Support Forecasting:** Provide data necessary for accurate space weather forecasts.

Key Components

The main components of space weather monitoring systems include:

- **Space-Based Observatories:** Satellites and spacecraft that monitor solar and interplanetary activity.
- **Ground-Based Observatories:** Instruments on Earth that measure geomagnetic and ionospheric conditions.
- **Data Processing and Analysis:** Systems and algorithms that analyze the collected data and generate forecasts and alerts.

15.2 Space-Based Observatories

Geostationary Operational Environmental Satellites (GOES)

The GOES series of satellites play a crucial role in space weather monitoring:

- **Instruments:** Equipped with X-ray sensors, solar imagers, and particle detectors to monitor solar activity.
- **Coverage:** Positioned in geostationary orbit, providing continuous monitoring of the Sun and near-Earth space.
- **Data Utilization:** Data from GOES satellites is used for real-time monitoring and forecasting of solar flares and CMEs (Space.com) .

Solar and Heliospheric Observatory (SOHO)

SOHO is a joint project by NASA and ESA focused on studying the Sun:

- **Instruments:** Includes the Extreme Ultraviolet Imaging Telescope (EIT), Large Angle and Spectrometric Coronagraph (LASCO), and others.
- **Mission:** Monitors solar wind, solar flares, and CMEs, providing early warnings of space weather events.
- **Achievements:** SOHO has significantly improved our understanding of solar dynamics and space weather phenomena .

Advanced Composition Explorer (ACE)

ACE provides real-time solar wind data critical for space weather forecasting:

- **Instruments:** Measures solar wind speed, density, temperature, and magnetic field.
- **Position:** Located at the L1 Lagrange point, providing a continuous view of the Sun-Earth line.
- **Role:** Data from ACE helps predict the arrival of CMEs and the onset of geomagnetic storms .

Parker Solar Probe

The Parker Solar Probe aims to study the Sun's outer corona:

- **Mission Goals:** Investigate the solar wind and the Sun's magnetic field to understand the origins of space weather.
- **Instruments:** Equipped with a suite of instruments to measure magnetic fields, plasma waves, and energetic particles.
- **Impact:** Provides unprecedented close-up observations of the Sun, improving our understanding of solar activity and space weather generation .

15.3 Ground-Based Observatories

Magnetometers

Magnetometers measure variations in Earth's magnetic field:

- **Global Network:** Deployed worldwide to monitor geomagnetic activity in real time.
- **Applications:** Data from magnetometers is used to detect and analyze geomagnetic storms and substorms.
- **Contributions:** Essential for understanding the interaction between solar wind and Earth's magnetosphere .

Ionosondes

Ionosondes measure ionospheric properties by sending radio waves and observing their reflection:

- **Function:** Determine electron density profiles and detect ionospheric disturbances.
- **Deployment:** Operated at multiple locations globally to provide comprehensive coverage.
- **Importance:** Critical for forecasting HF radio communication disruptions and understanding ionospheric behavior during space weather events .

Cosmic Ray Detectors

Cosmic ray detectors monitor high-energy particles from space:

- **Purpose:** Detect cosmic rays and solar energetic particles (SEPs) that can affect technology and human health.
- **Networks:** Examples include the Neutron Monitor Database (NMDB) and Muon Network.
- **Impact:** Data from these detectors helps assess radiation risks for aviation and space missions .

15.4 Data Processing and Analysis

Supercomputing and Data Analysis

Advanced computing is essential for processing the vast amounts of data collected:

- **Supercomputers:** Used to run high-resolution simulations and models of space weather phenomena.
- **Big Data Analytics:** Techniques such as machine learning and data mining are employed to analyze large datasets and improve forecast accuracy (Space.com).

Modeling and Simulation

Modeling and simulation play a crucial role in understanding and predicting space weather:

- **MHD Models:** Simulate the behavior of plasma in the Sun's atmosphere and solar wind.
- **Solar Wind Models:** Predict the arrival of solar wind structures at Earth.
- **Hybrid Models:** Combine empirical, numerical, and AI-based approaches to enhance forecast accuracy (Space.com).

Integrated Forecasting Systems

Integrated systems combine data from multiple sources to provide comprehensive space weather forecasts:

- **Real-Time Data Integration:** Systems like the Space Weather Modeling Framework (SWMF) integrate data from satellites, ground-based observatories, and models.
- **Forecast Generation:** Generate alerts and forecasts for various space weather phenomena, including solar flares, CMEs, and geomagnetic storms (Space.com).

15.5 Early Warning Systems

Purpose of Early Warning Systems

Early warning systems are designed to provide timely alerts about upcoming space weather events:

- **Alerts and Warnings:** Issued for solar flares, CMEs, and geomagnetic storms to enable preventive measures.
- **Communication Channels:** Information is disseminated through websites, email alerts, mobile apps, and other platforms.

Impact Assessment and Preparedness

Early warning systems help assess the potential impacts of space weather and prepare appropriate responses:

- **Risk Assessment:** Evaluates the likelihood and severity of impacts on critical infrastructure.
- **Preparedness Measures:** Implements actions such as adjusting satellite orbits, rerouting flights, and managing power grid loads (Space.com) .

Case Studies

The March 1989 Geomagnetic Storm

- **Forecasting Success:** Early warnings helped mitigate impacts on power grids and communication systems.
- **Impact:** Despite early warnings, the storm caused significant damage, highlighting the need for improved forecasting and mitigation strategies.

The Halloween Storms (2003)

- **Forecasting Improvements:** Enhanced forecasting methods and real-time data networks provided timely alerts for the Halloween storms.
- **Impact and Response:** Preventive measures were implemented, reducing the overall impact of the storms (Space.com).

15.6 Future Directions in Space Weather Monitoring

Advancements in Observational Technologies

Future advancements will focus on improving observational capabilities:

- **Next-Generation Satellites:** Launching new satellites with advanced sensors to monitor solar activity and space weather in real time.
- **Global Observing Networks:** Expanding global networks to improve data coverage and enhance forecasting accuracy.

Integration of Emerging Technologies

Emerging technologies will play a crucial role in advancing space weather monitoring:

- **Quantum Computing:** Utilizing quantum computing to process large datasets and run complex models more efficiently.
- **Internet of Things (IoT):** Integrating IoT devices for real-time monitoring of space weather impacts on critical infrastructure.

International Collaboration

Strengthening international collaboration will be essential for improving space weather monitoring:

- **Data Sharing Agreements:** Establishing agreements for sharing space weather data and forecasts among countries and organizations.
- **Joint Research Initiatives:** Collaborating on research projects to develop advanced monitoring techniques and improve global resilience (Space.com) .

Space weather monitoring systems are vital for detecting, forecasting, and mitigating the impacts of space weather on technology and human activities. This chapter has provided a comprehensive overview of the various components, technologies, and methodologies involved in space weather monitoring systems, highlighting their importance in ensuring the resilience of modern society.

1. **Space-Based Observatories:** Satellites and spacecraft monitor solar and interplanetary activity, providing critical data for space weather forecasting.
2. **Ground-Based Observatories:** Instruments on Earth measure geomagnetic and ionospheric conditions, complementing space-based data.
3. **Data Processing and Analysis:** Advanced computing techniques and integrated forecasting systems analyze collected data to generate accurate forecasts.
4. **Early Warning Systems:** Provide timely alerts and enable preventive measures to mitigate the impacts of space weather.
5. **Future Directions:** Advancing observational technologies, integrating emerging technologies, and strengthening international collaboration will enhance space weather monitoring capabilities.

This comprehensive chapter on space weather monitoring systems sets the stage for understanding subsequent chapters, which will explore specific space weather phenomena, their impacts on technology, and additional strategies for preparedness and response.

CHAPTER 16: INTERNATIONAL COOPERATION IN SPACE WEATHER RESEARCH

pace weather phenomena, such as solar flares, coronal mass ejections (CMEs), and geomagnetic storms, have global impacts on technology, infrastructure, and human activities. Addressing these challenges requires a coordinated international effort. This chapter delves into the importance, mechanisms, and achievements of international cooperation in space weather research.

16.1 The Importance of International Cooperation

Global Impact of Space Weather

Space weather events can affect multiple regions simultaneously, making international collaboration essential:

- **Communication Disruptions:** HF radio and satellite communication systems worldwide can be disrupted.
- **Navigation Systems:** Global Navigation Satellite Systems (GNSS) like GPS, GLONASS, and Galileo are affected by ionospheric disturbances.
- **Power Grids:** Geomagnetic storms can induce currents in power grids across continents.

Shared Resources and Expertise

Pooling resources and expertise enhances the ability to monitor and predict space weather:

- **Data Sharing:** International data sharing improves the accuracy of space weather models and forecasts.
- **Technological Development:** Collaborative development of advanced monitoring instruments and satellites.
- **Knowledge Exchange:** Sharing research findings and methodologies accelerates scientific progress.

16.2 Key Organizations and Agreements

International Space Environment Service (ISES)

ISES is a collaborative network of space weather service providers from around the world:

- **Founding and Purpose:** Established in 1962, ISES aims to share space weather data and forecasts.
- **Members:** Includes space weather centers from the United States, Europe, Japan, China, Russia, and other countries.
- **Activities:** Provides real-time space weather information, forecasts, and alerts to support global users.

World Meteorological Organization (WMO)

The WMO coordinates international efforts in weather, climate, and water-related activities, including space weather:

- **Space Weather Programme:** Established to integrate space weather observations into global meteorological efforts.
- **Collaboration:** Works with national meteorological and space agencies to enhance space weather monitoring and forecasting capabilities.

Committee on Space Research (COSPAR)

COSPAR promotes international cooperation in space research, including space weather:

- **Scientific Assemblies:** Hosts biennial assemblies to facilitate the exchange of research findings.
- **Publications:** Publishes the journal "Advances in Space Research" to disseminate scientific knowledge.

United Nations Committee on the Peaceful Uses of Outer Space (COPUOS)

COPUOS addresses international space policy, including space weather cooperation:

- **Working Group on Long-term Sustainability:** Focuses on the sustainability of space activities, including space weather monitoring and mitigation.
- **Guidelines:** Develops guidelines for enhancing the safety and resilience of space operations.

16.3 Collaborative Research Initiatives

International Heliophysical Year (IHY)

The IHY, celebrated in 2007-2008, aimed to advance the understanding of heliophysical processes:

- **Objectives:** Fostered international collaboration in space weather research and observations.
- **Achievements:** Led to the deployment of new monitoring instruments and enhanced data sharing among participating countries.

Space Weather Follow-On (SWFO) Program

The SWFO program, led by NOAA and international partners, focuses on improving space weather forecasting:

- **Mission:** To deploy advanced satellites and instruments for continuous space weather monitoring.
- **Partnerships:** Collaborates with ESA, JAXA, and other space agencies to share data and research findings.

International Living with a Star (ILWS)

ILWS promotes the study of the Sun-Earth connection and space weather impacts:

- **Goals:** Understand the Sun's influence on the Earth and heliosphere, and improve predictive models.
- **Activities:** Facilitates joint research projects, data sharing, and coordinated observations among member organizations.

16.4 Data Sharing and Integration

Global Data Networks

Establishing global data networks is crucial for comprehensive space weather monitoring:

- **ISES Real-time Data Network:** Provides real-time space weather data from multiple international sources.
- **SuperMAG:** An international collaboration that integrates magnetometer data from over 100 observatories worldwide.

Data Repositories

Centralized data repositories enhance accessibility and usability of space weather data:

- **NOAA's National Centers for Environmental Information (NCEI):** Archives and provides access to space weather data.
- **ESA's Space Weather Coordination Centre (SSCC):** Facilitates data exchange and coordination among European space weather providers.

Standardization of Data Formats

Standardizing data formats ensures compatibility and ease of use:

- **WMO's Space Weather Data Exchange:** Develops standardized data formats and protocols for international data sharing.
- **COSPAR's International Reference Atmosphere:** Provides standardized models for the Earth's atmosphere and ionosphere.

16.5 Capacity Building and Education

Training Programs

Training programs enhance the skills and knowledge of space weather researchers and forecasters:

- **ISES Workshops:** Conducts workshops and training sessions on space weather monitoring and forecasting techniques.
- **WMO Education and Training Programme:** Offers courses and resources on space weather for meteorologists and other professionals.

Educational Outreach

Educational outreach efforts raise awareness and interest in space weather research:

- **Public Lectures and Seminars:** Organizations like NASA and ESA host public lectures and seminars on space weather topics.
- **School Programs:** Initiatives like NASA's Space Weather Action Center engage students in space weather monitoring and research activities.

Research Fellowships and Scholarships

Research fellowships and scholarships support the development of future space weather scientists:

- **COSPAR Fellowships:** Provides funding for young scientists to conduct space weather research.
- **UN/Japan Long-term Fellowship Programme:** Offers scholarships for postgraduate studies in space science and technology, including space weather.

16.6 Achievements and Success Stories

Improved Forecast Accuracy

International collaboration has led to significant improvements in space weather forecast accuracy:

- **Enhanced Models:** Joint efforts have developed more accurate and reliable space weather models.
- **Timely Alerts:** Collaborative data sharing enables the issuance of timely alerts for space weather events.

Increased Resilience of Critical Infrastructure

Cooperative research and monitoring efforts have enhanced the resilience of critical infrastructure:

- **Power Grid Protection:** Improved forecasts help power grid operators implement protective measures during geomagnetic storms.
- **Aviation Safety:** Accurate predictions of solar radiation storms ensure the safety of aviation operations, especially on polar routes.

Advancements in Scientific Understanding

Collaborative research has advanced the scientific understanding of space weather phenomena:

- **Heliophysics Research:** International missions and studies have deepened knowledge of solar and heliospheric processes.
- **Geospace Interactions:** Joint projects have shed light on the complex interactions between the solar wind, magnetosphere, and ionosphere.

International cooperation in space weather research is essential for addressing the global challenges posed by space weather phenomena. This chapter has highlighted the importance, mechanisms, and achievements of international collaboration in this field, emphasizing the need for continued cooperation to enhance our understanding and mitigation of space weather impacts.

1. **Global Impact:** Space weather events have worldwide effects, necessitating international collaboration.
2. **Shared Resources:** Pooling resources and expertise enhances monitoring, forecasting, and research capabilities.
3. **Key Organizations:** Organizations like ISES, WMO, COSPAR, and COPUOS play crucial roles in facilitating international cooperation.
4. **Collaborative Initiatives:** Programs like IHY, SWFO, and ILWS promote joint research and data sharing.
5. **Data Integration:** Establishing global data networks and standardizing data formats improve accessibility and usability.
6. **Capacity Building:** Training programs, educational outreach, and research fellowships support the development of space weather scientists.
7. **Achievements:** International cooperation has led to improved forecast accuracy, increased resilience of critical infrastructure, and advancements in scientific understanding.

This comprehensive chapter on international cooperation in space weather research sets the stage for understanding subsequent chapters, which will explore specific space weather phenomena, their impacts on technology, and additional strategies for preparedness and response.

CHAPTER 17: FUTURE PROSPECTS AND CHALLENGES

As we move further into the 21st century, the field of space weather research and its applications face numerous prospects and challenges. This chapter explores these future developments, focusing on technological advancements, the evolving needs of society, and the scientific challenges that remain. With detailed information drawn from multiple sources, this chapter provides a comprehensive overview of what lies ahead for space weather research and its implications.

17.1 Technological Advancements

Next-Generation Satellites

The development and deployment of next-generation satellites will significantly enhance our ability to monitor and forecast space weather:

- **Enhanced Capabilities:** Future satellites will be equipped with more advanced sensors and instruments capable of providing higher resolution data and more comprehensive coverage of solar and geomagnetic activity.
- **International Collaboration:** Projects like the Space Weather Follow-On (SWFO) program by NOAA and partnerships with ESA, JAXA, and other space agencies will ensure a global network of advanced observational satellites (Space.com) .

Supercomputing and Data Processing

Advancements in supercomputing and data processing will be crucial for analyzing the vast amounts of data generated by space weather monitoring systems:

- **Increased Computational Power:** Utilizing supercomputers and quantum computing to run complex simulations and models of space weather phenomena.
- **Big Data Analytics:** Employing machine learning and artificial intelligence to analyze large datasets, improve forecast accuracy, and predict space weather events more reliably (Space.com) .

Miniaturized Sensors and Instruments

The miniaturization of sensors and instruments will facilitate the deployment of more comprehensive monitoring networks:

- **CubeSats and NanoSats:** Small, cost-effective satellites that can be launched in large numbers to provide detailed, real-time data on space weather conditions.
- **Distributed Sensor Networks:** Ground-based and space-based sensor networks that provide high-resolution data and improve the spatial coverage of space weather observations (Space.com).

17.2 Evolving Needs of Society

Increased Reliance on Technology

As society becomes increasingly reliant on technology, the importance of space weather research and mitigation strategies will grow:

- **Global Navigation Satellite Systems (GNSS):** The reliance on GNSS for navigation, timing, and communication will necessitate robust space weather monitoring and forecasting to ensure system reliability.
- **Power Grids:** Protecting power grids from geomagnetic storms will remain a critical priority, requiring continuous advancements in monitoring and mitigation technologies (Space.com) .

Space Exploration and Commercialization

The expansion of space exploration and the commercialization of space activities will drive the need for improved space weather forecasting:

- **Human Spaceflight:** Ensuring the safety of astronauts from radiation hazards during solar particle events (SPEs) and other space weather phenomena.
- **Commercial Satellites:** Protecting commercial satellites from space weather impacts to ensure the continuity of services such as communication, broadcasting, and Earth observation (Space.com) .

Climate Change and Environmental Monitoring

Space weather research will play a vital role in understanding the broader impacts of space weather on Earth's climate and environment:

- **Atmospheric Interactions:** Studying how space weather affects the upper atmosphere and its implications for climate models.
- **Environmental Monitoring:** Utilizing space weather data to improve the accuracy of environmental monitoring and forecasting systems (Space.com) .

17.3 Scientific Challenges

Understanding Fundamental Processes

Several fundamental processes in space weather science remain poorly understood, presenting significant challenges for researchers:

- **Solar Dynamics:** Gaining a deeper understanding of the processes that drive solar activity, including the formation and evolution of solar flares and CMEs.
- **Geospace Interactions:** Investigating the complex interactions between the solar wind, magnetosphere, ionosphere, and thermosphere during space weather events (Space.com) .

Modeling and Simulation

Developing accurate and reliable models of space weather phenomena is a major scientific challenge:

- **Multiscale Modeling:** Creating models that can simulate processes occurring at different spatial and temporal scales, from the solar surface to the Earth's magnetosphere.
- **Data Assimilation:** Integrating observational data into models to improve their accuracy and predictive capabilities (Space.com) .

Predictive Accuracy

Improving the predictive accuracy of space weather forecasts is a continuous challenge:

- **Lead Time:** Increasing the lead time of space weather forecasts to provide more time for preventive measures.
- **Uncertainty Quantification:** Quantifying the uncertainties in space weather forecasts to provide more reliable and actionable information (Space.com) .

17.4 Policy and Regulatory Challenges

International Collaboration

While international collaboration has significantly improved, further efforts are needed to enhance global coordination in space weather research and mitigation:

- **Data Sharing:** Establishing more comprehensive data-sharing agreements and protocols to ensure timely access to space weather data.
- **Joint Missions:** Coordinating international missions and research projects to leverage resources and expertise effectively (Space.com) .

Regulatory Frameworks

Developing and implementing regulatory frameworks for space weather mitigation is crucial for protecting critical infrastructure:

- **Standards and Guidelines:** Creating international standards and guidelines for space weather resilience in various sectors, including power grids and satellite operations.
- **Compliance and Enforcement:** Ensuring compliance with regulatory frameworks through monitoring and enforcement mechanisms (Space.com) .

Public Awareness and Education

Raising public awareness and education about space weather risks and mitigation strategies is essential for building a resilient society:

- **Outreach Programs:** Implementing outreach programs to inform the public, industry stakeholders, and policymakers about space weather impacts and preparedness measures.
- **Educational Initiatives:** Integrating space weather education into school curricula and professional training programs to develop a knowledgeable workforce (Space.com) .

17.5 Achievements and Success Stories

Improved Forecast Accuracy

International collaboration and technological advancements have led to significant improvements in space weather forecast accuracy:

- **Enhanced Models:** Development of more accurate and reliable space weather models through collaborative research.
- **Timely Alerts:** Implementation of real-time data networks and early warning systems that provide timely alerts for space weather events (Space.com) .

Increased Resilience of Critical Infrastructure

Efforts to enhance the resilience of critical infrastructure have yielded positive results:

- **Power Grid Protection:** Improved monitoring and forecasting capabilities help power grid operators implement protective measures during geomagnetic storms.
- **Aviation Safety:** Accurate predictions of solar radiation storms ensure the safety of aviation operations, particularly on polar routes (Space.com) .

Advancements in Scientific Understanding

Collaborative research has advanced the scientific understanding of space weather phenomena:

- **Heliophysics Research:** International missions and studies have deepened knowledge of solar and heliospheric processes.
- **Geospace Interactions:** Joint projects have provided new insights into the complex interactions between the solar wind, magnetosphere, and ionosphere (Space.com) .

The future prospects and challenges in space weather research are shaped by technological advancements, evolving societal needs, and ongoing scientific inquiries. This chapter has provided a detailed overview of the key areas that will define the future of space weather research, emphasizing the importance of international cooperation, technological innovation, and public awareness in addressing the global impacts of space weather.

1. **Technological Advancements:** Next-generation satellites, supercomputing, and miniaturized sensors will enhance space weather monitoring and forecasting capabilities.

2. **Evolving Societal Needs:** Increased reliance on technology, space exploration, and environmental monitoring drive the need for improved space weather research and mitigation.
3. **Scientific Challenges:** Understanding fundamental processes, improving modeling and simulation, and enhancing predictive accuracy are ongoing scientific challenges.
4. **Policy and Regulatory Challenges:** International collaboration, regulatory frameworks, and public awareness are critical for building resilience against space weather impacts.
5. **Achievements and Success Stories:** Improved forecast accuracy, increased resilience of critical infrastructure, and advancements in scientific understanding demonstrate the progress made through international cooperation and technological innovation.

This comprehensive chapter on future prospects and challenges sets the stage for understanding subsequent chapters, which will explore specific space weather phenomena, their impacts on technology, and additional strategies for preparedness and response.

CHAPTER 18: CASE STUDIES OF MAJOR SPACE WEATHER EVENTS

Space weather events have significantly impacted technology and human activities over the years. Understanding these events through detailed case studies helps us comprehend their causes, effects, and the measures taken to mitigate their impacts. This chapter presents comprehensive case studies of major space weather events, analyzing their causes, effects, and the lessons learned.

18.1 The Carrington Event (1859)

Overview

The Carrington Event is the most powerful geomagnetic storm on record, named after British astronomer Richard Carrington, who observed the solar flare that caused it:

- **Date:** September 1-2, 1859
- **Solar Activity:** A massive solar flare and coronal mass ejection (CME) were observed, leading to a geomagnetic storm on Earth.

Causes

- **Solar Flare:** The event began with a bright white-light solar flare observed by Carrington and another British astronomer, Richard Hodgson.
- **Coronal Mass Ejection:** The flare was followed by a CME that traveled directly toward Earth, arriving in just 17.6 hours, which is significantly faster than usual.

Effects

- **Auroras:** Bright auroras were visible around the world, even at low latitudes near the equator.
- **Telegraph Systems:** Telegraph systems, the primary communication technology at the time, experienced widespread failures. Telegraph operators reported sparks from their equipment, and some telegraph lines continued to operate without batteries due to induced currents.

Lessons Learned

- **Vulnerability of Technology:** The Carrington Event highlighted the potential vulnerability of technological systems to space weather, even with the relatively simple technology of the 19th century.
- **Preparedness:** Modern society must be prepared for similar events, which could have far more devastating effects on today's advanced technological infrastructure.

18.2 The March 1989 Geomagnetic Storm

Overview

The March 1989 geomagnetic storm is one of the most significant space weather events in recent history:

- **Date:** March 13-14, 1989
- **Solar Activity:** Caused by a CME following a significant solar flare.

Causes

- **Solar Flare:** An X15-class solar flare occurred on March 6, 1989, followed by a CME directed toward Earth.
- **Geomagnetic Storm:** The CME impacted Earth's magnetosphere, causing a severe geomagnetic storm.

Effects

- **Quebec Blackout:** The storm induced strong geomagnetic currents that caused the collapse of the Hydro-Québec power grid, leading to a 9-hour blackout affecting over 6 million people.
- **Satellite Malfunctions:** Several satellites experienced malfunctions, including disruptions in their control systems.
- **Communication Disruptions:** Shortwave radio communication was severely affected, and the auroras were visible as far south as Texas.

Lessons Learned

- **Power Grid Vulnerability:** Highlighted the vulnerability of power grids to geomagnetic storms and the importance of implementing protective measures.
- **Satellite Design:** Emphasized the need for robust satellite design to withstand space weather effects.

18.3 The Halloween Storms (2003)

Overview

The Halloween Storms were a series of intense solar storms that occurred in late October and early November 2003:

- **Date**: October 28 - November 4, 2003
- **Solar Activity**: Multiple powerful solar flares, including an X17.2 flare, one of the strongest ever recorded.

Causes

- **Multiple Solar Flares**: A sequence of powerful solar flares and associated CMEs impacted Earth over several days.
- **Rapid Succession of CMEs**: The close timing of the CMEs led to compounded effects on Earth's magnetosphere.

Effects

- **Satellite Damage**: Several satellites were damaged or experienced operational anomalies, including the loss of the Japanese ADEOS-2 satellite.
- **Power Grid Disturbances**: Geomagnetic storms induced currents that affected power systems, causing voltage instability in Sweden and South Africa.
- **GNSS Disruptions**: Significant errors in GPS positioning affected aviation and maritime navigation.
- **Auroras**: Auroras were visible at unusually low latitudes, including southern Europe and the southern United States.

Lessons Learned

- **Infrastructure Protection**: Reinforced the need for protective measures in power grids and communication systems.
- **GNSS Resilience**: Highlighted the importance of improving the resilience of GNSS to space weather events.

18.4 The Bastille Day Event (2000)

Overview

The Bastille Day Event was a major space weather event that occurred on July 14, 2000:

- **Date:** July 14-16, 2000
- **Solar Activity:** Triggered by an X5.7-class solar flare and an associated CME.

Causes

- **Solar Flare:** A powerful solar flare on July 14, followed by a CME.
- **Geomagnetic Storm:** The CME impacted Earth, causing a severe geomagnetic storm.

Effects

- **Satellite Anomalies:** Several satellites experienced anomalies and operational disruptions.
- **Radio Communication:** HF radio communications were disrupted, affecting aviation and maritime operations.
- **Auroras:** Auroras were visible at lower latitudes, including parts of the United States and Europe.

Lessons Learned

- **Spacecraft Shielding:** Emphasized the need for enhanced shielding and robust design of spacecraft to withstand space weather effects.
- **Communication Systems:** Reinforced the importance of resilient communication systems to maintain operations during geomagnetic storms.

18.5 The 2003 Halloween Storms

Overview

The Halloween Storms of 2003 were a series of intense solar storms that provided a wealth of data for space weather researchers:

- **Date:** October 28 - November 4, 2003
- **Solar Activity:** Included one of the most powerful solar flares ever recorded, an X17.2 flare.

Causes

- **Multiple Solar Flares:** A series of powerful flares and CMEs impacted Earth in quick succession.
- **Compounded Effects:** The close timing of CMEs caused compounded geomagnetic storm effects.

Effects

- **Satellite Failures:** Several satellites experienced severe anomalies, and some were rendered inoperable.
- **Power Grid Issues:** Geomagnetically induced currents caused power grid disturbances in Sweden and South Africa.
- **GNSS Disruptions:** Significant positioning errors affected GPS and other GNSS systems.
- **Auroras:** Auroras were seen at unusually low latitudes, as far south as Texas.

Lessons Learned

- **Infrastructure Resilience:** Highlighted the need for better protection of power grids and communication networks.
- **Research Collaboration:** Showcased the importance of international collaboration in space weather research and data sharing.

18.6 The St. Patrick's Day Storm (2015)

Overview

The St. Patrick's Day Storm was a significant geomagnetic storm that occurred in March 2015:

- **Date:** March 17-18, 2015
- **Solar Activity:** Triggered by a CME following an M-class solar flare.

Causes

- **Solar Flare and CME:** An M-class solar flare on March 15, followed by a CME that impacted Earth two days later.
- **Geomagnetic Storm:** The CME caused a G4 (severe) geomagnetic storm.

Effects

- **Auroras:** Auroras were visible at low latitudes, including parts of the United States and Europe.
- **GNSS Disruptions:** Significant errors in GPS positioning affected various applications, including aviation and maritime navigation.
- **Power Grid Impacts:** Minor impacts on power grids were reported, but no major outages occurred.

Lessons Learned

- **Forecasting Accuracy:** Highlighted the importance of accurate space weather forecasting to mitigate impacts.
- **GNSS Resilience:** Emphasized the need for improving the resilience of GNSS systems to space weather events.

18.7 The August 1972 Solar Storm

Overview

The August 1972 solar storm was a significant event that occurred during the Apollo program:

- **Date:** August 2-11, 1972
- **Solar Activity:** Featured a series of solar flares and CMEs.

Causes

- **Solar Flares:** Multiple solar flares, including an X-class flare on August 4.
- **CMEs:** Several CMEs followed the flares, impacting Earth and causing geomagnetic storms.

Effects

- **Satellite Anomalies:** Several satellites experienced operational disruptions.
- **Communication Disruptions:** HF radio communication was significantly affected.
- **Power Grids:** Minor impacts on power grids were reported.

Lessons Learned

- **Astronaut Safety:** Highlighted the potential risks of solar radiation storms to astronauts, leading to improved safety protocols for space missions.
- **Communication Systems:** Emphasized the need for resilient communication systems to maintain operations during geomagnetic storms.

The detailed case studies of major space weather events provide valuable insights into the causes, effects, and mitigation strategies for space weather phenomena. These events underscore the importance of continuous monitoring, accurate forecasting, and international collaboration in addressing the challenges posed by space weather.

1. **Vulnerability of Technology:** Space weather events can significantly impact technological systems, including power grids, satellites, and communication networks.
2. **Preparedness Measures:** Implementing robust monitoring, forecasting, and mitigation measures is crucial for minimizing the impacts of space weather events.
3. **International Collaboration:** Collaboration among international space weather agencies and organizations is essential for effective monitoring and response.
4. **Scientific Understanding:** Continued research and advancements in space weather science are necessary for improving our understanding and predictive capabilities.

Key Learnings from Major Space Weather Events

1. **Infrastructure Vulnerability:** The Carrington Event and the March 1989 geomagnetic storm highlighted the susceptibility of power grids and communication systems to space weather impacts. As technological dependency increases, so does the need for resilient infrastructure.
2. **International Collaboration:** Events like the Halloween Storms (2003) and the St. Patrick's Day Storm (2015) demonstrate the necessity of international data sharing and collaborative research efforts to enhance global preparedness and response capabilities.
3. **Technological Advancements:** Continuous improvements in monitoring and forecasting technologies, such as those deployed in the SWFO program and the use of AI in data analysis, are crucial for mitigating the impacts of space weather events.
4. **Preparedness and Mitigation:** Historical events underscore the importance of having robust preparedness measures and mitigation

strategies in place, such as enhanced shielding for satellites, protective measures for power grids, and accurate forecasting to provide timely warnings.

Case Study Summaries

18.8 The 1967 Solar Storm and the Cold War

Overview

The 1967 solar storm is notable for its impact during the Cold War, nearly causing a military conflict:

- **Date:** May 1967
- **Solar Activity:** Included a series of solar flares and a CME.

Causes

- **Solar Flares:** Several large solar flares, including a major X-class flare.
- **CME:** A CME associated with the flares impacted Earth, causing a geomagnetic storm.

Effects

- **Radar Interference:** The geomagnetic storm caused radar jamming over the polar regions, which the U.S. military initially mistook for Soviet interference.
- **Aviation and Communication:** Significant disruptions to aviation communication and navigation systems.

Lessons Learned

- **Military Preparedness:** The event highlighted the need for military awareness of space weather impacts to avoid misinterpretation during geopolitical tensions.
- **Communication Systems:** Reinforced the importance of resilient communication systems in both civilian and military contexts.

18.9 The February 2011 Solar Storm

Overview

The February 2011 solar storm was a significant event in the current solar cycle:

- **Date:** February 14-15, 2011
- **Solar Activity:** Triggered by an X2.2-class solar flare, the first X-class flare in over four years.

Causes

- **Solar Flare:** An X2.2-class flare, followed by a CME directed towards Earth.
- **Geomagnetic Storm:** The CME caused a moderate geomagnetic storm.

Effects

- **Communication Disruptions:** HF radio blackouts in China and other regions, affecting aviation communication.
- **Satellite Anomalies:** Temporary operational issues in some satellites.

Lessons Learned

- **Forecasting Improvements:** Demonstrated the need for continual improvements in space weather forecasting models to provide accurate predictions.
- **Technology Resilience:** Highlighted the ongoing need to enhance the resilience of communication and satellite systems.

18.10 The September 2017 Solar Storm

Overview

The September 2017 solar storm was marked by several powerful solar flares:

- **Date:** September 4-10, 2017
- **Solar Activity:** Included an X9.3-class flare, the most powerful in over a decade.

Causes

- **Multiple Solar Flares:** A series of large flares, including the X9.3-class flare, followed by CMEs.
- **Geomagnetic Storms:** The CMEs caused several geomagnetic storms, including a G4-class storm.

Effects

- **GNSS Disruptions:** Significant disruptions to GPS and other GNSS services, affecting navigation and timing.
- **Satellite Issues:** Temporary anomalies and increased radiation levels affecting satellite operations.
- **Auroras:** Visible at low latitudes, including the southern United States and Europe.

Lessons Learned

- **System Redundancy:** Emphasized the need for redundancy in GNSS and satellite systems to ensure continuous operation during space weather events.
- **Public Awareness:** Highlighted the importance of public awareness and education about space weather risks and mitigation measures.

The detailed case studies of major space weather events provide valuable insights into the causes, effects, and mitigation strategies for space weather phenomena. These events underscore the importance of continuous monitoring, accurate forecasting, and international collaboration in addressing the challenges posed by space weather.

1. **Infrastructure Vulnerability:** Major space weather events can significantly impact technological systems, highlighting the need for robust infrastructure and preparedness measures.
2. **International Collaboration:** Collaboration among international space weather agencies and organizations is essential for effective monitoring and response.
3. **Scientific Understanding:** Continued research and advancements in space weather science are necessary for improving our understanding and predictive capabilities.

18.11 The July 2012 Solar Storm

Overview

The July 2012 solar storm, though it missed Earth, had the potential to be one of the most catastrophic space weather events in modern history:

- **Date:** July 23, 2012
- **Solar Activity:** Characterized by an intense CME.

Causes

- **CME:** A massive CME, propelled by a solar flare, ejected towards Earth's orbit. However, Earth was on the other side of the Sun at the time, sparing it from the direct hit.

Effects

- **Near Miss:** Had the CME occurred just a week earlier, it would have directly impacted Earth, potentially causing widespread technological disruptions.
- **Potential Impacts:** Estimated that a direct hit would have caused severe geomagnetic storms, disrupting power grids, communication networks, and satellite operations globally.

Lessons Learned

- **Risk Awareness:** Highlighted the significant risk posed by extreme space weather events and the need for robust preparedness plans.
- **Improved Monitoring:** Emphasized the importance of continuous solar monitoring to provide early warnings for future events.

18.12 The February 1956 Geomagnetic Storm

Overview

The February 1956 geomagnetic storm was notable for its intensity and impact during an era with limited space weather monitoring capabilities:

- **Date:** February 23, 1956
- **Solar Activity:** Included a powerful solar flare and subsequent CME.

Causes

- **Solar Flare:** A significant solar flare followed by a CME directed at Earth.
- **Geomagnetic Storm:** The CME caused one of the most intense geomagnetic storms of the 20th century.

Effects

- **Communication Disruptions:** HF radio communications were severely affected, impacting military and civilian operations.

- **Auroras:** Auroras were visible at low latitudes, including parts of the Mediterranean and the southern United States.
- **Radiation Exposure:** Increased cosmic radiation affected aviation routes, particularly those over polar regions.

Lessons Learned

- **Early Awareness:** The event underscored the importance of early warning systems and continuous monitoring of solar activity.
- **Research Initiatives:** Led to increased research into geomagnetic storms and their potential impacts on modern technology.

18.13 The April 2001 Geomagnetic Storm

Overview

The April 2001 geomagnetic storm was a significant event that affected various technological systems:

- **Date:** April 11-12, 2001
- **Solar Activity:** Triggered by an X20-class solar flare and associated CME.

Causes

- **Solar Flare:** An X20-class solar flare, one of the most powerful recorded, followed by a fast-moving CME.
- **Geomagnetic Storm:** The CME impacted Earth, causing a severe geomagnetic storm.

Effects

- **Satellite Anomalies:** Several satellites experienced operational issues, including communication disruptions and orientation problems.
- **Power Grid Disturbances:** Geomagnetic induced currents affected power systems, causing minor power outages and operational disturbances.
- **GNSS Errors:** Significant positioning errors affected GPS accuracy, impacting navigation and timing services.

Lessons Learned

- **Satellite Resilience:** Emphasized the need for robust satellite design and enhanced shielding to withstand space weather impacts.
- **Improved Forecasting:** Highlighted the importance of accurate and timely space weather forecasts to mitigate the effects on critical infrastructure.

18.14 The January 2005 Solar Storm

Overview

The January 2005 solar storm was notable for its rapid onset and widespread effects:

- **Date:** January 20, 2005
- **Solar Activity:** Included an X7.1-class solar flare and a fast-moving CME.

Causes

- **Solar Flare:** An X7.1-class solar flare, followed by a CME traveling at nearly 3,000 kilometers per second.
- **Geomagnetic Storm:** The CME caused a severe geomagnetic storm upon impacting Earth's magnetosphere.

Effects

- **Communication Blackouts:** HF radio communications were severely disrupted, particularly affecting aviation and maritime operations.
- **Satellite Malfunctions:** Several satellites reported operational anomalies due to increased radiation levels.
- **GNSS Disruptions:** GPS and other GNSS systems experienced significant errors, affecting navigation services.

Lessons Learned

- **Rapid Response:** The event underscored the need for rapid response protocols to address space weather impacts quickly.
- **Public Awareness:** Highlighted the importance of public awareness and preparedness for space weather events.

18.15 The March 1991 Geomagnetic Storm

Overview

The March 1991 geomagnetic storm was a significant event that provided valuable data for space weather research:

- **Date:** March 24, 1991
- **Solar Activity:** Triggered by an X10-class solar flare and associated CME.

Causes

- **Solar Flare:** An X10-class solar flare, followed by a CME directed at Earth.
- **Geomagnetic Storm:** The CME caused a severe geomagnetic storm upon impacting Earth's magnetosphere.

Effects

- **Satellite Anomalies:** Several satellites experienced operational disruptions, including problems with communication and orientation.
- **Power Grid Issues:** Geomagnetic induced currents affected power systems, causing voltage instability and minor outages.
- **GNSS Errors:** GPS accuracy was significantly affected, impacting navigation services.

Lessons Learned

- **Infrastructure Vulnerability:** Highlighted the vulnerability of technological infrastructure to space weather events.
- **Research Collaboration:** Emphasized the importance of international research collaboration to understand and mitigate space weather impacts.

18.16 The August 2018 Solar Storm

Overview

The August 2018 solar storm was notable for its widespread impact and the valuable data it provided for future research:

- **Date:** August 26-27, 2018
- **Solar Activity:** Included an M-class solar flare and associated CME.

Causes

- **Solar Flare:** An M-class solar flare, followed by a CME that impacted Earth.
- **Geomagnetic Storm:** The CME caused a moderate geomagnetic storm upon impacting Earth's magnetosphere.

Effects

- **Communication Disruptions:** HF radio communications were affected, particularly impacting aviation operations.
- **Satellite Anomalies:** Several satellites experienced minor operational issues due to increased radiation levels.
- **GNSS Errors:** GPS and other GNSS systems reported positioning errors, affecting navigation services.

Lessons Learned

- **Enhanced Monitoring:** Highlighted the importance of continuous monitoring and real-time data analysis for accurate forecasting.
- **Public Education:** Emphasized the need for public education on space weather risks and preparedness measures.

The detailed case studies of major space weather events provide valuable insights into the causes, effects, and mitigation strategies for space weather phenomena. These events underscore the importance of continuous monitoring, accurate forecasting, and international collaboration in addressing the challenges posed by space weather.

1. **Infrastructure Vulnerability:** Major space weather events can significantly impact technological systems, highlighting the need for robust infrastructure and preparedness measures.
2. **International Collaboration:** Collaboration among international space weather agencies and organizations is essential for effective monitoring and response.
3. **Scientific Understanding:** Continued research and advancements in space weather science are necessary for improving our understanding and predictive capabilities.

This comprehensive chapter on case studies of major space weather events sets the stage for understanding subsequent chapters, which will explore specific space weather phenomena, their impacts on technology, and additional strategies for preparedness and respo

CHAPTER 19: THE ROLE OF EDUCATION AND PUBLIC AWARENESS

This book has delved into the multifaceted domain of space weather, elucidating its physics, effects, and the critical measures needed for mitigation. From the foundational understanding of the sun's influence and the dynamic interactions within the magnetosphere to the intricate effects on communication systems, power grids, and satellite operations, the comprehensive coverage provides insights into the significance of space weather on modern technology and infrastructure. Through detailed case studies and advancements in forecasting methods, this book has underscored the necessity of continuous monitoring, international collaboration, and technological innovation in mitigating the impacts of space weather.

19.1 Introduction to Education and Public Awareness

Importance of Education

Education plays a pivotal role in enhancing public understanding of space weather phenomena and their implications:

- **Knowledge Dissemination:** Educating the public about space weather helps demystify complex scientific concepts and fosters a deeper appreciation of the subject.
- **Preparedness:** Informed individuals are better prepared to respond to space weather events, mitigating potential impacts on daily life and critical infrastructure.

Role of Public Awareness

Public awareness initiatives are essential for:

- **Risk Mitigation:** Raising awareness about the risks associated with space weather can prompt proactive measures at individual, community, and governmental levels.
- **Support for Research and Policy:** An informed public is more likely to support funding for space weather research and the development of protective policies and technologies.

19.2 Educational Strategies

Formal Education

Incorporating space weather into formal education curricula is crucial:

- **School Programs:** Integrating space weather topics into science curricula at primary and secondary school levels to build foundational knowledge.
- **University Courses:** Offering specialized courses in space weather and related fields at the university level to train the next generation of scientists and engineers.

Informal Education

Informal education initiatives can reach a broader audience:

- **Public Lectures and Seminars:** Hosting talks by experts in space weather to educate and engage the public.
- **Workshops and Hands-On Activities:** Organizing workshops and interactive sessions to provide practical insights into space weather phenomena.

19.3 Public Awareness Campaigns

Media and Communication

Leveraging media and communication channels to raise public awareness:

- **Documentaries and TV Shows:** Producing educational documentaries and TV shows focused on space weather and its impacts.
- **Social Media Campaigns:** Utilizing social media platforms to disseminate information and engage with a wider audience.

Outreach Programs

Implementing outreach programs to educate communities:

- **Community Engagement:** Conducting outreach programs in local communities to spread awareness and provide educational resources.
- **Collaborations with Educational Institutions:** Partnering with schools, universities, and museums to facilitate outreach activities and exhibitions.

19.4 Case Studies of Successful Education and Awareness Programs

NASA's Space Weather Action Center

NASA's Space Weather Action Center (SWAC) is a notable example of an effective educational initiative:

- **Interactive Learning:** Provides interactive online tools and resources for students to learn about space weather.
- **Educational Modules:** Offers educational modules and activities designed to engage students and teachers in the study of space weather.

ESA's Space Weather Coordination Centre (SSCC)

ESA's SSCC plays a crucial role in public education and awareness:

- **Public Outreach:** Conducts public outreach activities, including exhibitions, public lectures, and educational materials.
- **Educational Resources:** Provides comprehensive educational resources on space weather for teachers and students.

NOAA's Space Weather Prediction Center (SWPC)

NOAA's SWPC is instrumental in raising public awareness:

- **Public Alerts:** Issues public alerts and warnings about space weather events, providing timely information to help mitigate impacts.
- **Educational Materials:** Develops and distributes educational materials to inform the public about space weather and its effects.

18.5 Challenges in Education and Public Awareness

Complexity of Space Weather Concepts

One of the main challenges in educating the public about space weather is the complexity of the subject:

- **Simplification:** Finding ways to simplify complex scientific concepts without losing accuracy is crucial for effective education.
- **Engagement:** Keeping the public engaged and interested in a highly technical subject requires innovative educational approaches.

Limited Resources and Funding

Limited resources and funding can hinder education and public awareness efforts:

- **Funding for Educational Programs:** Securing adequate funding for educational programs and outreach activities is essential.
- **Resource Development:** Developing high-quality educational materials and resources requires significant investment.

18.6 Future Directions in Education and Public Awareness

Integrating Advanced Technologies

Leveraging advanced technologies to enhance education and public awareness:

- **Virtual Reality (VR) and Augmented Reality (AR):** Utilizing VR and AR technologies to create immersive educational experiences about space weather.
- **Online Platforms:** Expanding the use of online platforms and digital tools to reach a global audience.

Collaboration and Partnership

Fostering collaboration and partnerships to amplify educational efforts:

- **International Cooperation:** Collaborating with international organizations and educational institutions to share resources and best practices.
- **Public-Private Partnerships:** Engaging private sector partners in educational initiatives to leverage additional resources and expertise.

Sustained Public Engagement

Ensuring sustained public engagement through continuous efforts:

- **Regular Updates:** Providing regular updates and new educational content to keep the public informed and engaged.
- **Interactive Initiatives:** Developing interactive initiatives and competitions to maintain public interest and participation.

The role of education and public awareness in space weather research and mitigation cannot be overstated. By educating the public and raising awareness about the risks and impacts of space weather, we can foster a more informed and prepared society. Through formal and informal education, public awareness campaigns, and innovative approaches, we can enhance our collective resilience to space weather phenomena.

1. **Educational Strategies:** Implementing comprehensive educational strategies, including formal and informal education, is crucial for building foundational knowledge about space weather.
2. **Public Awareness Campaigns:** Utilizing media, communication channels, and outreach programs to raise public awareness and engage communities.
3. **Challenges and Solutions:** Addressing challenges such as the complexity of space weather concepts and limited resources through innovative educational approaches and collaboration.
4. **Future Directions:** Leveraging advanced technologies, fostering collaboration, and ensuring sustained public engagement to enhance education and awareness efforts.

This comprehensive chapter on the role of education and public awareness provides a fitting conclusion to the book, emphasizing the importance of informed and engaged societies in mitigating the impacts of space weather.

"THE END"

ABOUT THE AUTHOR

Jennifer Marlercurtis & Klaus Scherer

Jennifer Marlercurtis is an esteemed researcher and author in the field of space weather. With a background in astrophysics and a passion for education, Jennifer has dedicated her career to advancing our understanding of solar and space phenomena. She has been instrumental in numerous international collaborative projects aimed at enhancing space weather forecasting and mitigation strategies. Jennifer is known for her ability to communicate complex scientific concepts to diverse audiences, making significant contributions to both academic research and public awareness.

Klaus Scherer is a distinguished professor and scientist specializing in heliophysics and space weather research. With a prolific career spanning several decades, Klaus has authored numerous influential publications on the interactions between solar activity and the Earth's magnetosphere. His work has been pivotal in the development of advanced models and forecasting techniques for space weather phenomena. Klaus is highly regarded for his commitment to fostering international cooperation in space weather research, contributing to global initiatives that enhance our preparedness for space weather events.

Together, Jennifer Marlercurtis and Klaus Scherer bring a wealth of knowledge and expertise to the field of space weather research, making significant strides in both scientific understanding and public education. Their collaborative efforts have resulted in groundbreaking research and innovative approaches to mitigating the impacts of space weather on modern technology and society.

Made in the USA
Las Vegas, NV
16 February 2025

18229705R00085